W9-CFL-384

Flying & Learning:

Basics

For Every Pilot

WITHDRAWN
BEATRICE, NEBR. 68310

Cessna 152

Piper Tomahawk

WILLIAMW

Flying & Learning:
Basics
For Every Pilot

By William P. Heitman

Foreword by Wolfgang Langewiesche

Dreamflyer Publications

© 1997 William P. Heitman
All rights reserved.

No part of this book may be reproduced or transmitted in any form or by any means, electronic, mechanical, including photocopying, recording, or by any information storage and retrieval system, except in the case of reviews, without the express written permission of the publisher, except where permitted by law.

Every effort has been made to ensure that permission for all pertinent material was obtained. In addition, every effort has been made to ensure that all material within this book is accurate, up to date, and acknowledged where necessary. Any material not formally acknowledged or changes due to governmental policy will be corrected and/or included in all future editions of this work subsequent to notification by any appropriate sources.

Should anyone find small errors in this book, please consider relaying them to the publisher. Constructive comments are also welcome.

ISBN 0-9660156-0-6

Cover art by Jerry Miller, a semi-retired architectural and aircraft illustrator from Cary, North Carolina.

Original illustration on page 109 by Ryan C. Teasley.

Photos on title page and back cover by Gary Whitford.

Dreamflyer Publications
P.O. Box 11583
Durham, N.C. 27703 U.S.A.

E-mail: Drmflyrpub@aol.com

Manufactured in the United States of America.
First edition.

Dedication

This book is dedicated
to every person
who has had the courage
to strap on an airplane
with me.

Table of Contents

Acknowledgments

I thank all the Certified Flight Instructors (CFIs), pilot examiners, FAA inspectors, and pilots of various experience who have passed along tips, reminders, explanations, and memory aids. Many of these ideas are included in some way in this work, passed along to you, with hope that it helps widen your understanding of aviation. There are some instructors who merit special recognition:

Douglas E. Moore (Doug), whom I still call "Coach," was my first flight instructor. More than once I know that he rolled his eyes in doubt and grasped his seat in fear.

Paul A. Craig, whose book knowledge rivals that of the best pilots, was my beloved tormentor during the course of my aviation college degree. A few times he pushed me and the plane to the edge of our respective envelopes, and I learned from it.

John E. Benton, the "Commander," frequently knew which of my strings to pull. Frequently, he pulled them. John was a proof-reader.

Jerry D. Weathers, the "Bossman," is the hardest working individual that I have ever met. He teaches by example. Jerry was a proof-reader.

B. Winfield Causey (Winn) is a rare individual who by his born nature combines the qualities of teacher, gentleman, businessman, and professional all in the right proportions. I could not have been more fortunate than to land at his airfield.

Thomas N. Jones and Lawrence F. Lambert are FAA Safety Program Managers who are liked and respected by all who know them. I hope the FAA knows and understands how dedicated and valuable these two gentlemen are to the industry they serve. Tom and Larry were proof-readers.

John Jenkins is a former Designated Pilot Examiner (DPE), current flight instructor, and a good fellow. John was a proof-reader.

Charlie Causey is the master A&P mechanic. If he signs off an airplane as airworthy, I will not think twice about flying it. That is, after a good pre-flight. Charlie was a proof-reader.

Barrie S. Davis is a P-51 Mustang Ace from the 325th Fighter Group in World War II. He is retired from the printing and publishing business and came into my life at just the right time with just the right knowledge and skills to help with the final editing and formatting.

There are many others to whom I am thankful. I will quickly mention some of these, but this list is by no means all inclusive: "Chip" Davis, Don Culp, John McLain, Bruce Fogleman, Kieran O'Dwyer, Alan Brooks, Marrow Smith, Bob Frederick, Byron Smith, Bruce Marshall, "Butch" Davis, Johnny Henley, Reuben Edwards, Ron Troutman, Barbara Baron, Linwood Roberts, Bill De Brauwer, Wolfgang Langewiesche, Richard Bach, Jerry Miller, Ryan Teasley, Mark R. Bero, Craig Greenlaw, and Gary Whitford. Craig and Gary are two of my private pilot success stories.

Craig and his wife are responsible for lighting the initial fire under my feet that led to this book.

Gary was invaluable in giving his time to help finish this book. He took many of the photos and was a great proof-reader.

Last but certainly not least I would like to acknowledge the support given by my wife, Cathy. She is not a pilot, but she does wear a large halo for her patience with my passion. She claims that she actually learned something about flying and pilot responsibilities while proof-reading the manuscript. She was an absolutely brutal but priceless proof reader, forcing me to explain everything. Thank you, honey, I love you.

They all have done their jobs well. I hope in turn that I have done mine as well.

William P. Heitman

Foreword

What I liked in this book was the emphasis on very simple basic things. The student comes to the airplane and to flight instruction as ignorant as a small child comes to nursery school. The one learns to tie his shoelaces; the other learns to follow a straight line marked on the ground. A good nursery school benefits its graduates for many years in their further scholastic careers. More really primitive primary flight instruction would benefit its students similarly. In our present flight instruction we disregard those initial problems. Phil Heitman does not. That's what I liked about his book.

Wolfgang Langewiesche, April 1996
Author of *Stick and Rudder*

Introduction

The Bridge Builder

An old man going a lone highway,
Came, at the evening cold and gray,
To a chasm vast and deep and wide,
The old man crossed in the twilight dim,
The sullen stream had no fear for him;
But he turned when safe on the other side
And built a bridge to span the tide.

Old man, said a fellow pilgrim near,
You are wasting your strength with building here;
Your journey will end with the ending day,
You never again will pass this way;
You've crossed the chasm, deep and wide,
Why build this bridge at evening tide?

The builder lifted his old gray head;
Good friend, in the path I have come, he said,
There followed after me to-day
A youth whose feet must pass this way.
This chasm that has been as naught to me
To that fair-haired youth may a pitfall be;
He, too, must cross in the twilight dim;
Good friend, I am building this bridge for him!

—Anonymous

I found this poem in a book of poetry that had a copyright of 1931. It is a shame that we do not know the author of this poem, because I think every conscientious flight instructor would agree that this author has captured the essence of the passion—and compassion—of teaching the gift of flight. Since even before the Wright brothers, flight for man has been a constant building of ideas and experiences and subsequent sharing of that knowledge. If it were not for those who have built their bridges back to us who follow, human flying prowess would not have progressed as rapidly and as safely as it has.

When I first started as a flight instructor, I quickly learned that I needed in reserve at least three explanations for every maneuver and performance standard. I learned that although two students were on the exact same lesson, flying back to back on the same day, a clear explanation that made a positive connection with the first student might not, even when repeated word for word, connect with the next student. This difference was because people with varied backgrounds learn different things at different rates and, as I know now, with different explanations.

Instead of allowing this situation to become frustrating, I viewed each question as a personal challenge to find the explanation that best fit each person. I researched different texts and books and asked other instructors for examples, styles, and techniques that seemed to work. I incorporated these into my own regimen. As a result, not only did my students benefit, but I learned and grew as a pilot as well. I have always encouraged my students to ask lots of questions. If it were one that I could answer, they benefited. If it were a question that I could not answer, I was off to the books to find the answer for them and for myself. We both learned. We both benefited.

I have found over the years that there are common problem areas in learning to fly, such as those faced by a student pilot learning to land, that are elusive to teach and to learn because of their dynamic nature. How can an instructor teach someone to feel the airplane as it sinks into ground effect, especially when you recognize that every landing is new and essentially different from the last? The answer is that he or she cannot. The instructor can only put that student in that position repeatedly to allow the student to acquire the feel while the instructor acts as a safety net and counselor. Once we all were intimidated by a flying machine, and each of us reacted with similar impulsive mechanical inputs rather than smooth balanced ones. Balance is acquired through practice.

But then, if every landing is different, how does one standardize the learn-to-land process? The answer again is that, except for the practical test standards and certain flight training syllabi, one cannot. While giving biennial flight reviews or checking pilots out in a new aircraft, I have seen many different traffic pattern techniques exercised. Each ultimately accomplishes the same result of putting the tires on the runway, but some of these techniques are good and some are horrendous. Most techniques can be traced back to the original flight instructor. If that instructor were conscientious, proficient, and methodical, it will show. If the instructor were apathetic and just there to build time, his or her attitude usually will rear its ugly head later through the actions of the resultant pilot. Keep in mind that there may be a large difference between a pilot's and an instructor's techniques and their procedures. While technique is non-regulatory, certain procedures may very well be.

The general premise of this book is to take the pilot, whether he or she be a student, private, commercial, CFI, or even a person yet to start on the flight training journey, from pre-flight of a non-complex single engine airplane, through take-off, then a flight around the standard traffic pattern of a non-towered (sometimes mistakenly called uncontrolled) airport, and finally, through the process of a landing. Sounds simple enough. Almost too simple. But, as you will see, the basics are what are overlooked most frequently by many pilots. A simple trip around the pattern will exercise many basic flight skills when it is flown correctly. It is unfortunate that too many pilots get their certificate(s) and never really practice the basics, or, more specifically, never practice the traffic pattern again. They take all the factors involved in the process for granted. Unfortunately, taking anything in aviation for granted is inviting trouble.

Another reason for the straightforward and basic approach of this text is that many student pilots try vainly to tell their instructors that all they really need is landing practice. Hopefully, this book will illustrate that there is more to the traffic pattern than only practicing landings. In fact, the whole traffic pattern sequence consists of various smaller pieces of a puzzle that have to be practiced first, piece by piece, outside of the pattern. Climbing and descending turns, airspeed transition, checklist usage, airspeed control, division of attention, wind drift correction, flight at slower air speeds, aircraft reconfiguration, altitude and distance judgment for the wind conditions—these are just some of the fundamentals that must be learned by a pre-solo student in the practice area before the skills are brought back into the traffic pattern. It is in the traffic pattern where the pieces should all snugly fit together. Getting into the traffic pattern too soon will frustrate the student, because he or she has never mastered all of the pieces of the puzzle.

It is sad that we still have accidents in the traffic pattern. One accident is too many. More alarming is that many of these accidents occur while under Visual Flight Rules (VFR) conditions at non-towered airports. Perhaps something read herein will prevent someone from making that cross-controlled stall when turning from base to final. Or perhaps someone else might look a little harder during the preflight of an airplane for problems which can arise. That, in and of itself, would make this endeavor and all the time and effort spent on it very worthwhile. For the private pilot who cannot seem to find the time or money to visit the airport as often as he or she needs to stay proficient, that person can read and use this book as a vehicle to mentally "chair fly" an airplane so that his or her proficiency can become, at the very least, a more comfortable state of mind.

Especially for the CFI/student relationship, this book can serve as the baseline from which common ground lines of communication can be established. In other situations, it may provide the bridge or extra explanation needed to eliminate a student's frustration. Perhaps it may even inspire some other student/applicant to continue his or her training to a higher level. Even so, this book is not flight training gospel. It is my treatise. It should be used as a learning tool and to promote ideas for safer flying.

Flight maneuvers, cross country navigation, and instrument flying techniques are entirely different topics. The techniques, examples, photographs and illustrations that I will present are intended to stimulate pilots into thinking about how they can improve their flying, starting with the basics. The learning process first must begin with the thinking process. In this day and age of complicated airspace and numerous regulations, it is vital that we have thinking pilots in our skies.

What we do as pilots is special. It takes special training, special knowledge, special preparation, and a special thought process. It is the job of the CFI to nurture that thought process and promote the type of understanding where correlations can be made. The student's duty is to inquire constantly, gain insight, practice the basics, and continue to hunger for flight and aircraft knowledge. It is my hope that by relaying in this book the aviation basics that I have learned through education, experience, and instructing, I can help provide a learning window, my bridge, for anyone who comes after me and has that hunger for aviation, airplane, and overall flight knowledge.

Every pilot is a student, every pilot must think, and every pilot should be learning.

—William Phillip Heitman

What Do You Need to Be a Pilot?

"A pilot must possess the innate faculty of selective and instinctive discrimination of stimuli of the sensory motor apparatus to harmoniously adjust metabolic changes in physiological and psychological equilibrium in such manner to comprehend and assimilate instruction in the attributes essential to perform the intricate and complex operations which constitute the details of pilotage."

—*South African Pilot* magazine

CHAPTER 1

Weather

Aerology: The study of the properties of the air and of the atmosphere.

There are so many good books on weather that I will not discuss the topic extensively here. What I will advise is that every flight should begin with a pre-flight briefing from a Flight Service Station (FSS), even if it is only a local training flight.

I remember one instance when I had a student scheduled and, before the student arrived, my early call to the FSS told of an approaching cold front. Our runway was 6/24, and the wind was light, steady, and straight down 24. The front was approaching from the northwest. I had been briefed that it was moving fast, but I knew we had some time to fly. Our usual practice area was to the northwest, so one option would have been to go there. If we felt increasing turbulence we could turn and easily outrun the front back to the airport. Two other less used practice areas were to the east and southeast, but these were not options because by the time we would have felt a change in the wind, the front would already have passed the airport. I did not want to subject a pre-solo student to a possibly very bumpy lesson.

Being of conservative nature, I opted to stay in the traffic pattern. This decision suited the student just fine, because he was eager to practice his approaches and landings anyway. Besides, with the front being a fast mover, I was wary of it and, if necessary, I wanted to be able to get down quickly. This turned out to be one of my more fortunate instructor decisions, because the front had, unknown to me, turned into a squall line that was moving at 50 to 60 miles per hour.

After four or five trips around the pattern, I could begin to see in the distance a dark overhead shelf of clouds. When back on the ground after landing, I told the student that we would do one more trip around the field to a full stop and call it a day. We seemed to have plenty of time. Here is how I recall that last takeoff to landing.

Immediately after liftoff, the clouds still seemed much too distant to be of consequence for this last hop around the patch. There was blue sky overhead. Turning on the crosswind leg, the dark, high-shelf cloud movement was suddenly seen as approaching fast. After we turned downwind, the sky already was darkening as the cloud shelf quickly cut off the blue sky. Less than 5 minutes had passed since the decision to take off. At the beginning of the base leg all blue sky was gone from overhead, and the situation was looking and feeling ominous. I told the student to cut the approach short and head directly for the runway. At that time we began to feel the wind gusts pick up and shift to a right crosswind. I told my student that I might have to take the controls completely for this last landing. He quickly agreed.

The landing itself turned out to be a medium level crosswind one, with the wind gusty and from the right as anticipated. It did turn out to be too much of a crosswind for the student to handle by himself, but I permitted his hands to be on the controls with mine as we touched down. As we taxied in, the gentleman on the unicom called us and said that the airport Automated Weather Observation System (AWOS) was reporting the wind at 25 to 30 miles per hour directly across the runway. Then I realized that if we had been only seconds longer in our approach, the crosswind could easily have exceeded the maximum crosswind landing capability of our Piper Tomahawk. Luck had been on our side, but I felt uneasy with the close call. The small trainer continued to be rocked by gusts as we taxied to the ramp. Quickly tying the plane down, we were walking to the terminal building when the bottom fell out with torrents of wild diagonal rain driven by very strong winds. Had we gone to the northwest practice area, I am sure that we could not have turned and out-run this dangerous squall line. It was moving much too fast, and this whole situation turned out to be one that I am grateful to be here to talk about.

I learned some strong lessons here. I knew that the cold front was moving fast, but I did not call to get an update from Flight Service. This was my job. If I had re-called the FSS and been advised that a squall line had developed, then this particular flight lesson would have been conducted as a ground session. I knew squall lines were dangerous; I know it even better now. I knew that the weather could change quickly; I know it even better now.

Every lesson should be an opportunity for the student (and the CFI) to practice his or her conversation with an FAA weather briefer. If the Fixed Base Operation (FBO) has a speaker phone, the CFI should use it with the student so that he or she can hear how a weather briefing is conducted. On subsequent occasions the CFI should listen on the speaker phone or on another line to evaluate how a student handles his or her own briefing so that the briefing can be critiqued and improved. Then the instructor should ask the student some other weather-related questions. Everyone should develop the habit of thinking about the weather and checking for current information every time before blasting off into the wild blue yonder. If there is a delay in your departure, check again with the FSS before proceding. More important, every pilot should get the habit of learning as much as he or she can about the element through which we fly. Get a good book on aviation weather and read it twice.

"A superior pilot uses his superior judgment to avoid those situations which require the use of his superior skill."

—Frank Borman, former Astronaut

CHAPTER 2

Weight & Balance

Aeronautics: The art or science of flight.

Weight and balance is another one of those subjects on which I do not think it necessary to elaborate in this text. However, this does not diminish the importance of this subject. What I will do is describe how weight and balance are essential to safe flight. It is assumed that every pilot will not exceed the weight and balance limitations for the aircraft which he or she will fly, once properly trained. Every pilot should be intimate with the weight and balance requirements for each aircraft flown. Primary flight training is the venue for learning weight and balance, one on one, with the instructor. Before I had soloed, I learned this rule: "If you're out of weight and balance, you're dead."

The importance and drama of weight and balance was fully illustrated when my instructor showed me, at my request, the charred wreckage of a four-place aircraft in which four people died. They crashed through a manufacturing plant's roof, where wreckage came to rest suspended in the roof beams. Some, perhaps all, were alive as the aircraft dangled precariously. Plant workers were the last to hear the screams as the wreckage burst into flames. A sad but true real life drama.

Here is the scenario later told to me by an eyewitness. The four adults arrived at the airport one Friday evening around dusk, excited about their weekend trip to the beach. The pilot had called earlier and instructed line personnel to fill the plane's fuel tanks full. The group drove next to the airplane, transferred their luggage from their car to the plane, and, after one parked the car, they all boarded the aircraft. It is not known whether a pre-flight was conducted. After takeoff, the towered field controller instructed the pilot to make a turn (which is not unusual) and, during the turn, the aircraft fell from the sky.

I am not an accident investigator, but there are numerous red flags associated with this accident. That an aircraft designer would not or could not design a four place aircraft that could accommodate four adults, their luggage, tanks full of fuel, and still remain at maximum gross weight or less, seems incongruous, but it is true. In fact, most four place aircraft cannot accommodate this load. Center of gravity or CG (the balance) is another issue, but giving the benefit of doubt, let us assume that this particular aircraft, even with its heavy load, was within the center of gravity limits.

The takeoff roll that evening was probably longer than usual, because of excess weight, but the runway was long enough to allow the horsepower and propeller efficiency to pull that weight up to flying speed. If the pilot held the best rate of climb speed, this most likely produced a shallower than usual climb-out, but the wings were faithful in lifting all that weight into the air. What the pilot did next was to turn at the controller's request. No one knows how quickly the pilot responded or how steeply he banked, but keep in mind that this probably was a climbing turn with full power. What we do know is that during the turn, something went wrong. My assessment is that the pilot allowed the aircraft wing to stall.

Early in training, a fledgling pilot learns about and practices stalls. One of the stalls that he or she learns and practices is one during a full power, climbing turn. Stall maneuvers are practiced at a safe altitude and always with less than gross weight. This pilot had to do them, or he would never have obtained his certificate; therefore this fact in itself is not enough to cause alarm. But stall speed is a function of weight. Even with wings level, more weight requires a higher angle of attack for the wing to generate the lift needed to raise that weight. But the angle of attack can be increased only so much before the wing will stall. It is important to know also that a given wing will always stall at the same angle of attack (see Figure 2-1).

There are several other things to discuss before tying all this information together. Load factor is also a function of weight. In fact, load factor is a weight multiplier, and it multiplies according to the angle of bank. This is called apparent weight. The higher the angle of bank, the

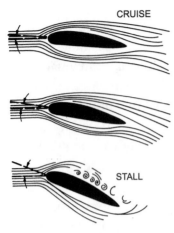

CRUISE

STALL

Figure 2-1
Airflow about a wing in various
angles of attack from cruise
to a stall

more the load factor and, thus, the more the wing will react as if it were carrying a heavier load. Given a constant indicated airspeed (such as best rate of climb speed), the speed at which an aircraft will stall will increase with weight and/or load factor. So, as apparent weight increases, the stall speed, normally much lower than climb-out speed, also increases until it reaches the narrow window of climb-out speed.

Here now is the scenario again: A pilot already in excess of maximum gross weight takes off. It is dark or near dark. The wing has to work harder with a higher than normal angle of attack to lift the extra weight, narrowing its margin to a stall angle. The pilot initiates a bank, which increases the apparent weight (load factor) held by the wing. When the pilot tries to hold his full power climb speed, all his margins are lost. The wing stalls, there is not enough altitude to initiate a recovery, and four people experience horror before their fiery deaths. Now, even though my own assessment of this accident is pure speculation, the explanation given is a plausible one. And as pilot-in-command (PIC) of the aircraft, this pilot was responsible for this flight, this accident and ultimately for the deaths of his friends.

The accident just described did not have to happen. Even though the pilot may have known his plane intimately, he pushed it beyond its limitations. All this could have been avoided if the total weight had been lowered to a weight not above the maximum gross weight allowed by the manufacturer. Something should have been left on the ground. The question is what. The answer is some fuel, some baggage, somebody, or some combination thereof.

One of the best examples illustrating weight and balance is comparing the Cessna 152 (C-152) and the Piper Tomahawk. Both aircraft have two seats and the same maximum gross weight of 1,670 pounds. This weight is the maximum allowed for their wings to burden into the air. Let us assume that the same two people, each at 170 pounds, are going to fly first the C-152 with full tanks (24.5 gallons) and then the Tomahawk with full tanks (30 gallons). Fuel weighs 6 pounds per gallon.

Using the sample basic empty weight from the C-152's "Weight and Balance" section of the pilot's operating handbook, we get 1,136 pounds for the aircraft plus 340 pounds for the people plus 147 pounds of fuel (24.5 x 6), for a total of 1,623 pounds. The pilot and passenger could carry 47 pounds of baggage and still be under maximum gross weight.

Now look at what happens when the same calculation is performed for the Tomahawk. Again using the sample basic empty weight from the Tomahawk's information manual "weight and balance" section, we add 1,169 pounds for the aircraft plus 340 pounds for the people plus 180 pounds of fuel (30 x 6), to equal a total of 1,689 pounds. The travelers are already 19 pounds over gross weight without any baggage! If these two people want to fly the Tomahawk safely within maximum gross weight, they will have to leave slightly more than 3 gallons of fuel on the ground. If they want to fly with the same 47 pounds of baggage that they carried in the C-152, they will have to now leave a total of 11 gallons of fuel on the ground.

I want to emphasize here that these numbers will differ from all other individual C-152s and Tomahawks. You must figure your aircraft's actual weight and balance using its own unique basic empty weight, your weight, your instructor's or passenger's weight and the aircraft's fuel capacity. My numbers are strictly empirical and are used only to illustrate a point.

When you are first learning to fly, take the time to fully know and understand what weight and balance really mean. Your instructor will be glad to help you. If you are a newly certified pilot, do not disregard weight and balance as something you had to muddle through just to get your "ticket." It is more important now than ever before, because your passengers will be trusting you with their lives. You are responsible as a pilot-in-command.

If you are a "seasoned" pilot, stop and think about the last time you

Weight and Balance for N_____

Station	Weight	Arm	Moment
Airplane			
Oil (7.5 lb/gal)			
Pilot			
Front Seat Passenger			
Second Row Passenger			
Third Row Passenger			
Third Row Passenger			
Front Baggage			
Baggage			
Aft Baggage			
Baggage			
Fuel*			
Fuel*			
Fuel*			
Misc			
Misc			
Totals			

Weight X Arm = Moment

$$\frac{\text{Total Moment}}{\text{Total Weight}} = \text{Center of Gravity (CG)} _____$$

Max Forward CG = _____

Max Aft CG = _____

Pressure Altitude = _____   Ambient Temperature = _____

Density Altitude = _____   Distance Required to Clear _____ = _____

*Gasoline = 6 lb/gal

Figure 2-2

worked through a weight and balance problem. If you cannot remember when that was, then it is time to work a weight and balance problem using the sample form provided (see Figure 2-2). If you are unmotivated or unsure, find an instructor to help you. Use the opportunity at the Biennial Flight Review (BFR) to become familiar with weight and balance again. A responsible pilot would never take off with a happy-go-lucky attitude or with question marks in his or her head. To do so will invite disaster.

"It is impossible to accurately measure the results of aviation safety. No one can count the fires that never start, the aborted takeoffs that do not occur, the engine failures and the forced landings that never take place. And one can neither evaluate the lives that are not lost, nor plumb the depths of human misery we have been spared. But the individuals with the flight controls, fueling hose, wrench, radar, or dispatch order can find lasting satisfaction in the knowledge that they have worked wisely and well, and that safety has been the prime consideration."

—Author unknown

CHAPTER 3

Pre-flight

"Know your airplane. Know it well. Know its limitations and, above all, know your own limitations."

—R.A. "Bob" Hoover, Airshow Pilot

If there were ever a truism spoken in aviation, it is that takeoffs are optional; landings are not. In most cases, the pre-flight check of the airplane is the only opportunity to find a safety discrepancy with the machine that you are about to take into the air. After the takeoff is not the time to find a problem (or have it find you), because then you may be landing sooner than scheduled. One big goal of safe flying is landing when and where you choose.

NASA's Aviation Safety Reporting System database finds that an omission occurring during pre-flight most likely will manifest itself before or during the takeoff phase. The causes of these pre-flight errors or oversights are mainly twofold. The first is the "hurry-up syndrome" caused by stress from a tight schedule and/or weather discomfort. When it is cold, raining, or both, there is strong temptation to abbreviate the pre-flight inspection.

The second reason people rush through pre-flight activities is a feeling of complacency that the pre-flight is a routine event. Some owners feel that if their aircraft were functional when they last left it, then it should be the same when they return to it. I call this method the blind faith pre-flight. In fact, most of these same people believe that all of flying is routine and if nothing has happened before, it will not happen now. Are you willing to bet your life on it? It takes strong discipline to repeatedly do things correctly. Perform the pre-flight each time as if your life depends on it. It does.

You must pre-flight yourself as well as the aircraft. Do not fly if you are not current, not qualified, or if you doubt your skills for the flight conditions. An acronym for a self pre-flight is "I'M SAFE." Ask yourself if you are suffering from:

- ✈ **I**llness ?
- ✈ **M**edication ?
- ✈ **S**tress ?
- ✈ **A**lcohol ?
- ✈ **F**atigue ?
- ✈ **E**motion ?

If the answer to any one of these is yes, stop and think seriously before you pilot an airplane, even if only around the traffic pattern. If the answer is yes to any medication or alcohol, you definitely should not act as pilot-in-command.

While the types of things that you check during pre-flight procedures on most light aircraft are generally the same, there are distinct differences between manufacturers' designs and equipment. There are even differences between models of the same manufacturer. The only way to be sure that you are pre-flighting the proper things on an aircraft is to follow the manufacturer's pre-flight checklist found in the Pilot's Operating Handbook (POH) for that aircraft. Get a POH for your aircraft and read it. It is full of valuable information. Unfortunately, however, the checklists in these POHs sometimes do not cover every detail required for a thorough pre-flight inspection. Checklists often need to be customized. Even among some make and model aircraft, equipment variations and unique cockpit modifications can lead to overlooked, sometimes critical items when a generic checklist is used. Also, the more functions an aircraft has, the more complex the checklist is likely to be. This book cannot realistically cover all the possible scenarios for which you must be on guard. A lifetime of flying and thousands of hours of experience will be your absolute best teaching resource.

It would be equally impossible in this book to cover the pre-flight regimen for every aircraft. Although there are many types of trainer aircraft in use today and new types appearing readily, I have no preference for any one model, design, or manufacturer.

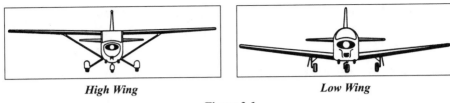

High Wing   Low Wing

Figure 3-1

Most examples that I cite, however, will come from the C-152 or the Piper Tomahawk. This is not an endorsement for these aircraft, but they do represent the most abundant trainers in use in America today. They also serve as examples for the basic differences between a high wing and low wing aircraft pre-flight (see Figure 3-1).

If you are undecided as to which type of aircraft you want to learn in, consider the following: Which type of aircraft will you likely fly or upgrade to after you obtain your certificate? Will you consider renting or buying a 4 place Piper, Cessna, Beechcraft, or the like? The transition from a high wing to high wing aircraft is relatively easier than a high wing to low wing aircraft. Nevertheless, the most important fact for you to know before reading further is that, above all else, I am trying to stimulate your thought processes about how to approach the pre-flight procedures of your aircraft or any aircraft. Always remember that actions on the ground are preferable to reactions in the air.

Inside the Cockpit

If you are an aircraft owner, you should know the paperwork that must be logged, available, and/or on board your plane before any flight. If you are a renter, then your pre-flight inspection might start before you step outside the FBO building. An aircraft used only for rental is required by law to have an annual inspection. Of course, maintenance is required throughout the year on those items which are flight mandatory. It is your right as a renter to see the maintenance records for that aircraft. If a rental operator is unwilling or unable to disclose these records, then your best course might be to find another operation.

If the aircraft is used in a commercial operation, such as flight instruction, it is required to have an inspection every 100 hours. This inspection is the same as is done for an annual inspection. The only difference in a 100 hour inspection and an annual is that an Airframe & Powerplant (A&P) mechanic can sign the logbooks for a 100 hour but only an A&P with an Inspection Authority (IA) can endorse an annual, although the same work is necessary for both. An annual can serve as the 100 hour inspection, but the 100 hour cannot serve as the annual. Only after you are satisfied that the proper inspections and maintenance have been completed should you proceed to pre-flight the aircraft of your choice.

After reaching the aircraft, the place to start any pre-flight is with a review of the overall appearance of the aircraft. There is an old saying in aviation: *If it looks good, it will fly good.* While not always true, this is a good rule of thumb. If you arrive at the aircraft, and there is oil puddled underneath or streaming out of the cowling, turn around, walk back to the FBO, and hand the operator the key. But if a tire were flat, this would not necessarily be a red flag which would deny you the chance to fly. Only you can decide. If the first impression is good, proceed to the cabin, but do not put the starter key into the ignition switch. Place it on the dash where you can see it at all times or keep it in your pocket. If the propeller needs to be turned through later in the pre-flight, knowing the location of that key is true peace of mind. A little safety backup to ensure engine non-operation is smart, especially if there is an inquisitive child in the cockpit waiting for you to finish your inspection outside.

Next you should turn to the "Normal Procedures" section of the POH and follow the checklist. However, not all checklists make reference to checking the documentation that is required in an aircraft for flight. Here is some help.

The acronym "AROW" is used to remind us of the requisite papers which should be inside the aircraft on any flight. The mnemonic aid used to be ARROW, with the second "R" representing the radio station license for the aircraft. However, the FCC no longer requires these. A clear plastic pouch is usually provided to house some or all of the necessary papers. The "A" stands for the airworthiness certificate. This document should always be positioned in front of the others and face out so that it can be seen clearly by any of the passengers or crew.

The "R" is for the registration of the aircraft. Just like your car registration, this is a paper record of the owner.

Some people think that the "O" stands for the pilot's operating handbook. This is not necessarily true. It more accurately represents the operating limitations of the aircraft or its operations and performance. While most limitations and performance standards will be found in the POH, some can also be found as placards in the cockpit. I suggest that you become knowledgeable with the contents and location of the POH and all placards in and around your aircraft.

The "W" represents weight and balance information. Always use the most current weight and balance for calculations. The weight and balance information is usually found in the Weight and Balance section of the POH. Previous non-current weight and balance information should have a diagonal line drawn through it and have the word superseded written along the line together with the superseded date.

It is the PIC's responsibility to check these documents and ensure that they are indeed on board the airplane. If they are not, and this aircraft is flown and subsequently the subject of an FAA ramp check, the FAA inspector will hold the PIC, not the aircraft owner, accountable and may issue a violation to the pilot. Also, do not forget that the pilot must have on his or her person flight and current medical certificates to be legal for PIC duty.

Most checklists agree that at this point in the inspection all in-cockpit flight control locks or restraints should be removed. There may be other restraints to remove later outside the aircraft, such as a rudder gust lock (see Photo 3-2). Most checklists agree also that the master electrical switch should be turned on at this time. What most do not tell you is that you should first check to ensure that all the avionics equipment

Photo 3-2
Rudder gust lock

has been turned off individually first. This action prevents an electrical surge or spike through this sensitive equipment when the master switch is turned on and when the engine is started. The Tomahawk POH does allude to this action on the last two pages in the very last section of this manual (section 10, entitled "Safety Tips"). There is some valuable information in this section, and I do not know why it appears at the end of the book. It is proof, though, that you should be familiar with every page of your POH. If there is an avionics master switch, turn it off to block any surges.

A reason for turning on the master switch is to check the fuel quantity indicators. Remember what they indicate so that you can verify gauge accuracy with the quantity that is actually observed in each tank later. This mental check will give you an indication as to how reliable the quantity indicators really are. This also is a good time to listen for the spinning "whirr" of the electrically-driven, turn coordinator gyroscope, if your aircraft is so equipped. Step outside the airplane and check to see that the anti-collision light(s) work. On a typical C-152, this would be a flashing or rotating beacon on top of the tail. On most Tomahawks, there is not a rotating beacon; rather there are strobe lights on the wing tips. While checking the strobes during a Tomahawk preflight, be sure to check the electric stall warning horn also on the leading edge of the left wing. Whichever anti-collision lighting system you have on your aircraft, make sure that it works. You may not be legal to fly unless it operates properly. For a night flight, be sure to check all of the lights on the aircraft, inside and outside the cockpit, for proper coloration and working order.

If your aircraft is equipped with electric flaps, put them down now, one notch at a time, looking for simultaneous operation. Once complete, turn the master switch off so that the flap actuating rods and hinges can be checked during the walk-around. Do not leave the master switch on or proceed to raise the electric flaps simply to check for flap up-operation. At times when the battery is already weak or during cold mornings, I have seen a battery nearly drained because of this practice. A flap motor can pull a lot of precious current from the battery, robbing its ability to start the engine. You should wait until the engine is running and the battery is being replenished by the alternator before checking the flaps for up-operation.

If the aircraft has manual flaps, lower one notch at a time and observe that each side lowers simultaneously, pausing only long enough to verify equal side-to-side operation and position. If the flaps on either side operate at different rates or if the flaps stop at unequal positions, this is called a split flap condition. It can cause seriously detrimental flight characteristics. Flying the base leg or final approach is not the time to find out that there is a split flap situation.

Some type of fuel sampling device should be located in the cockpit. There are different types, but all accomplish basically the same thing. They allow you to evaluate a fuel sample from the tanks for color, water, rust particles, dirt, gasket debris, bladder debris, or anything that would interrupt the flow of fuel to the engine (see Photo 3-3). I personally like the fuel sampler with the reversible screwdriver. This type permits either a Phillips head or flat head screw to be tightened where needed.

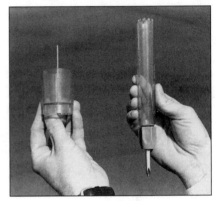

Photo 3-3
Two different fuel samplers

Outside the Aircraft

Some checklists suggest starting with one wing. Others teach starting with the empennage (tail group) or fuselage. Interestingly, the word empennage comes from the French language and translates into tail feathers, like the feathers on an arrow. The feathers give an arrow directional stability, and the empennage gives an airplane the same.

For the sake of descriptive consistency, I am going to start at the left side of the fuselage, work around the empennage to the right wing, nose, left wing, and back to the cockpit where it all began. The idea is to work your way around the aircraft in a flow pattern so that everything possible is checked while at a particular "station" and thereby greatly

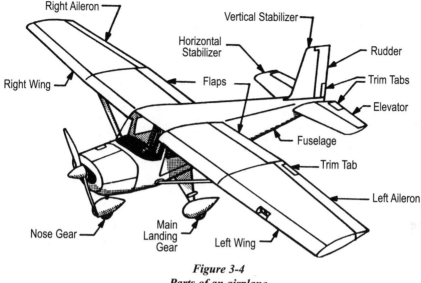

Figure 3-4
Parts of an airplane

reduce (but not rule out) the need to backtrack later. You can use your plane's checklist to define the "stations." No matter what direction your POH suggests for pre-flight, try going the other direction once in a while to add variety and to see things from a different perspective (see Figure 3-4).

Think of the fuselage as a soda can or a thin aluminum tube that can be wrinkled or buckled with stress. For an airplane, the stress can come from aerobatics or other maneuvers that are not approved for the aircraft or from an extremely hard landing. Signs of stress could show as wrinkled skin, popped or loose rivets or even missing rivets.

Look at the fuselage left side static air ports if they are present on that side of your airplane. The placement of static ports on aircraft is determined by aerodynamic design engineers. These small openings are always positioned 90° from the direction of flight in an area of relatively undisturbed airflow so that a constant sampling of atmospheric pressure can be accomplished. Static ports can become clogged from dirt, mud, wind-blown debris, an aircraft cleaning or waxing, and/or insects. Make sure they are clear of any obstruction but do not blow into them. Blowing into the ports can cause damage to pressure-sensitive instruments. Also try not to touch the static port opening(s), as oil from the fingers can subsequently attract dust and dirt.

Touch all inspection covers to verify that the screws are holding them tight. At the empennage there usually are two types of lock-nuts holding control surface hinges together. Regardless of whether there is a lock nut with a cotter pin through it or a self-locking nut, check each one for security. If you are of short stature and the locking nut cannot be seen, feel for it with your fingers. This would include each hinge of every control surface and also the connections where control cables attach to move a control surface. On self-locking types the shaft of the bolt should be felt protruding at least 2 or 3 threads from the end of the nut. Check the cables themselves for frayed strands. Do not forget to remove any external wind gust locks and be sure to untie the tail. I can speak from experience. You cannot taxi very well if you do not untie the tail. Neither will you soon forget the experience of having airport jokers watch your frustration and then reminding you of it later.

Check the control surfaces for freedom of movement and full deflection in both directions. If your airplane is approved for spins and you plan to practice spins, here is a valuable check for which you will need a partner. Have your partner get in the cockpit, deflect the yoke full forward, and hold it there. You must be at the elevator verifying that it is down. Then gently pull it up, using both hands, one on each side of the rudder. If there is more than one inch of slack, play, or travel in the trailing edge of the elevator, do not spin intentionally. Your ability to recover could be jeopardized because of sloppy control response. This check is accomplished easily on a C-152 but, because the Tomahawk is a "T" tailed airplane, you will need a ladder to accomplish it on this air-plane. Check for the same type of play with the rudder using full deflec-tion to both sides, but note that once again this check is easier on the C-152.

As you did with the fuselage, check the horizontal and vertical sta-bilizers, and the rudder and elevator, for any skin damage or stress. Scan and feel rivet heads for signs of a weak bond. If there are questions in your mind, get an instructor's or A&P's opinion and alert the aircraft owner of the problem. Finally, at and around the empennage, examine the remaining inspection plates for security.

You are now working up the right side of the fuselage. Here is a good opportunity to check the security of all the antenna bases where they attach to the fuselage. Losing an antenna is a good precursor to losing radio communication and/or navigation capabilities. If the aircraft is equipped with a static port on this side of the fuselage, as is true with the Tomahawk, check it now as you did the one before.

Now you should be at the right rear wing root. If you are pre-flighting a C-152, this is a good time to sample fuel from the right fuel tank sump. Get at least half a sampler full, more if necessary. If you are checking for the blue color of 100 Low Lead (100LL), do not hold the sample to the sky. Even water will look blue when held up against a blue sky. Instead, hold the sample up against the fuselage. Most fuselages are white or a light color, and you will be able to see clearly a color contrast and confirm that it is the right color. Another fuel that is acceptable for the C-152 is 100 octane, which is green in color.

Some other aircraft may be approved to use 80 to 87 octane which is a red color. Water held to the red hues of a dawn or dusk can be misleading and look like this color of fuel.

Check your POH to determine which fuel grade is acceptable in your airplane's engine. Do not be satisfied with color alone. Some non-aviation diesel fuels are now required to be dyed blue-green or red depending on their application. Home heating oils and kerosene which have a high sulfur content are dyed blue-green. This dye, depending on its combination, can sometimes look like 100LL or 100 octane. Low sulfur, non-highway diesel fuel is colored red. If your FBO uses a kerosene space heater or there is some other piece of equipment at your airport, such as a tractor, that uses diesel fuel, there could be a supply of this deceivingly correct-colored fuel available. Check the smell of the fuel sample from your aircraft. Gasoline and aviation octane fuels have a distinctly different smell than that of diesel fuel, Jet-A or kerosene.

Using another of your senses, check the feel of the fuel sample. Pour a little over your fingers and rub them together. Gasoline and aviation fuels will feel light and usually air dry quickly. They rob the fingers of body oils, leaving the fingers feeling dry. Kerosene, diesel and

Jet-A will feel heavy, oily, and will not dry quickly. They leave the fingers feeling greasy. If you suspect the fuel is wrong, do not fly the aircraft until you have investigated further. A reciprocating engine will not run on Jet-A, diesel, or kerosene. It will, however, sometimes run long enough on the fuel left in the fuel lines to get you into the air and to the scene of an accident.

While at this station, check the right flap for locknut/hinge security (see Photo 3-5). Grasp the flap at the trailing edge and try to move it up and down. It should move very little, if at all. To check the push rod that actuates the flap, use your thumb and forefinger. Rotate the rod around its longitudinal axis and feel the play. The play should be there because of the swivel joints on each end. The swivel joints are needed because the push rod does not travel straight back and forth. If the push rod does not rotate or have some play, it

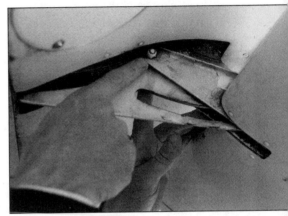

Photo 3-5
Check for locknut security

Photo 3-6
This bent push rod would be dangerous

may be bent or rusted, causing it to bind and preventing the flap from actuating properly. A faulty push rod condition might lead to a dangerous split flap situation in flight (see Photo 3-6).

Move to the right aileron now and check it for freedom of movement. Watch the yoke inside the cockpit as you move the aileron. When the aileron is moved up, the yoke should rotate toward you; aileron down, the yoke rotates away. This is especially important to check after maintenance. The mechanic is only human and could have reversed the cables after a refit, replacement, or inspection. Check all aileron and flap hinges for hinge pin security. Feel the aileron movement for binding or rubbing. During hot summer months, I have seen the aileron rub and bind with a wing tip fairing as a result of expansion from heat. Inspect the aileron push/actuator rod just as you did for the flap.

If you are pre-flighting a C-152, check the security of the aileron counterweights. These are lead weights that are riveted to the aileron. They are located on the front side of the aileron behind the main body of the wing. You may notice that they are attached with larger rivets than those used on the rest of the aileron. If a weight should come loose, it can bind the aileron, causing control problems. Many students look confused at this information and query me about the fact that they know the ailerons actually counterbalance each other. Then I talk aerodynamics for a moment.

There is a difference between a counterweight and a counterbalance. While the ailerons do counterbalance each other for the most part, the counterweights help to reduce flutter on each aileron by moving the center of mass closer to the pivot point. Flutter really is only a concern at high airspeeds, near and above redline, but during normal operation there would be little concern. The counterbalance action of one aileron going up and the other down does, however, balance the control pressure felt in the yoke during flight. On the empennage, the elevator and the rudder have counterweights and counterbalances as well.

Remember for a moment the action of pushing down on the elevator. Did you notice that the tips of the elevator moved up? Those are the elevator aerodynamic counterbalances.

In order to understand how aerodynamic counterbalances work, you need to imagine the aircraft moving forward with air flowing over the elevator. If the yoke is pushed forward, the elevator moves down. Air striking its underside tries to push it back up. The counterbalances also receive wind striking their underside, but because they are sticking up, the wind is trying to hold them up as well (see Photo 3-7).

This interaction helps to reduce the control pressure needed on the yoke to hold the elevator down. Being smaller in surface area, however, the counterbalances do not overpower the elevator. They are sometimes weighted inside to help with leverage. The elevator counterweights on the C-152 and Tomahawk are enclosed within these counterbalances.

Photo 3-7
Airflow strikes the counterbalance and the elevator

The rudder also has an aerodynamic counterbalance. It is usually at the top and works on the same principle as the counterbalances on the elevators.

Now, continuing with the pre-flight. At the wing tip fairing, look for any large or long cracks that might cause the fairing to peel back in flight, creating copious amounts of drag. In the same thought, we do not want the fairing to separate from the aircraft and strike anyone on the ground. The forces and stresses exerted on aircraft components during flight will cause cracks to worsen with time. Small cracks can sometimes be cured from creeping by a technique known as stop-drilling. Stop-drilling is simply drilling a 1/8 inch hole at the very tip of the crack. A qualified A&P mechanic must do this for you. Although not guaranteed, this drilling probably will relieve the stress and stop the crack from spreading (see Photo 3-8).

At this point during the pre-

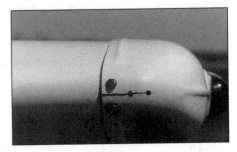

Photo 3-8
Stop-drilling to ease cracking

flight inspection, you should untie the right wing. While you are under the wing, check the fuel vents that are present. Like the static port, these can be clogged with dirt, mud, or insects seeking a refuge. A clogged fuel vent will cause a vacuum to be formed in the fuel tank, resulting in fuel flow stoppage. One way of explaining a vacuum in a fuel tank is the analogy of pouring fuel from a gas can into a lawnmower. You must open a vent on the opposite side of the gas can to allow air to replace the vacated fuel volume or flow is interrupted. If the vent is not opened, the gas usually can get some air back up through the spout causing another swap of gas for air, but the aircraft fuel tank cannot function in this manner.

Another way of explaining a vacuum in a fuel tank is to think about the operation of a syringe. Pulling the plunger allows outside air to replace the volume, displaced by the plunger, through the needle end. Normal use permits the plunger (representing the fuel) to move (or flow) up and down the barrel easily. If, however, you put your finger over the end where the needle would attach and pull the plunger, the plunger will move easily until a vacuum is formed and then resist moving further. No matter which example helps you understand, always remember that a vacuum in a fuel tank can easily result in an unplanned landing. The fuel vents on the Tomahawk are small tubes sticking straight down with a forward facing bevel cut. They are located near where each landing gear strut meets the wing.

While you are at this station, inspect the right tire for proper inflation and tread wear. Look for bald or flat spots or irregular wear on the tire (see Photo 3-9). These may indicate that the tire consistently stops rotating at the same spot after takeoff, meaning that it will touch down on the same spot each time and cause localized wear. This type of wear on a nose tire can lead to nose wheel shimmy when landing. Because of this

Photo 3-9
Irregular tire wear

tire wear inspection, I prefer to train pilots in airplanes which are not equipped with wheel pants. Wheel pants do not allow the access needed to illustrate and explain potential tire problems. Be sure to remove wheel chocks from around the right tire (see Photo 3-10).

Photo 3-10
PVC tubing wheel chocks

Neither the C-152 nor the Tomahawk have oleo struts as part of their main gear assembly, but if your aircraft has an oleo strut (see Photo 3-11) for this right main gear, wipe it clean periodically. This helps extend the life of the seals, which can be cut and scored from dirt particles. Now look at the brakes. The pad(s) should be at least as thick as a paper match is wide. Any fluid puddle under the assembly is a dead giveaway that something is wrong. Even if fluid is not obvious, feel the brake fluid tubing connection. A small droplet still hanging there may mean that the reservoir already is completely drained.

Photo 3-11
Main gear oleo strut

I once landed an older Piper Cherokee that had the single hand brake bar, which actuated both main brakes, and only the left brake held. It was interesting having to apply right rudder to keep the plane tracking straight. Thank goodness for long runways. Try to prevent these unusual situations from happening to you.

If you are pre-flighting a Tomahawk, this is a good time to drain the right fuel sump (see Photo 3-12) and check the fuel quality as described before. While the thought of checking fuel is fresh on your

Photo 3-12
Fuel sump drain

mind, stand up and visually inspect the fuel quantity. Do not trust what you saw earlier on the fuel gauges inside the cockpit. Your eyes are the only thing you can truly trust in this inspection. If there is one thing that is (or should be) common to all powered aircraft pilots, it is love for petroleum hydrocarbons. In the same thought, there absolutely is no excuse for running out of fuel in an aircraft. Staying on the subject of fuel, here is a quick review of the five things that fuel does for the air-craft: it causes combustion, cools, lubricates, cleans (solvent), and affects weight and balance.

On the C-152, it is now time to get up and check the fuel on this high wing aircraft. Regardless of whether you need a ladder or whether your aircraft already is equipped with footsteps and handholds, get up there. The first thing to look for is a vented fuel cap.

Almost all Cessnas are required by an Airworthiness Directive (AD) to have a vented fuel cap on the right side. If it does not, inade-quate fuel flow or fuel flow stoppage from the right side can occur because of a vacuum forming in that tank. Be sure to check visually for adequate fuel quantity and scan the top of the wing for structural dam-age before you get down. Structural damage on the wing, like any on the fuselage, often is apparent as airframe skin wrinkling and/or rivet prob-lems. At dusk or during darkness, it is nearly impossible to see into the

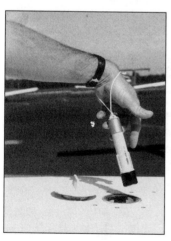

Photo 3-13
Flashlight security loop

tank to check fuel quantity. Since a flash-light is mandatory for a night pre-flight, I recommend finding and using one that has a hole at the base. A string can be tied through the hole to make a loop which can be extra insurance that the unthinkable incident of dropping the flashlight into the fuel tank does not happen (see Photo 3-13).

Now you should be standing at the right forward wing root. Here is an opportu-nity to examine the right side fresh air and ventilation intakes. On the C-152, one intake is on the leading edge of the wing root. Air comes through this vent into the cockpit overhead of the right seat. A mirror image intake is located on the left wing, routing fresh air to the left seat. There are two additional ram air intakes on the right side fuselage of the C-152. The first is a small protruding scoop just below the windshield. Air is routed from this intake to behind the instrument panel and cools heat produced by avionics. There is a similar scoop on the left side of the fuselage serving the same purpose. The last fresh air intake on the right side of the C-152 is a small rectangular flap which is governed by a push-pull cable and knob in the cockpit. Air coming through the flap can be used for cooling, or it can be mixed with heated air to regulate cockpit temperature (see Photo 3-14).

Photo 3-14
C-152 fuselage-to-cockpit fresh air intake

On the Tomahawk, the right side fresh air inlet is a recessed NASA/NACA scoop design just above the leading edge of the wing root (see Photo 3-15). Check especially for bees nests in all air intakes. As you accelerate down the runway for takeoff, bees left in the intakes may be trapped inside and could force their way into the cockpit helped by the increasing ram air. I assure you that, if this happens, no living creature inside the cockpit will be happy. Creatures outside the cockpit may get a display of

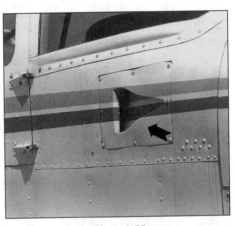

Photo 3-15
Tomahawk right side cabin fresh air intake

unexpected low level aerobatics soon after your takeoff.

I must make a point here. It is important that you know the precise reason for each and every hole in your aircraft. If you come across a hole during pre-flight inspection which looks as if it were designed to be there, find out what its purpose is. I remind you of the quote by Bob Hoover at the beginning of this section: "Know your airplane. Know it well."

At this juncture there is another distinct difference between the C-152 and the Tomahawk. Though these two aircraft have essentially the same engine, access to the engine compartment is considerably better on the Tomahawk. I like the Tomahawk's access because it permits me to better see and reach any bird or bee nests and to remove them. One type of nest to search for is the mud or dirt dauber's nest. Some people call these insects black wasps. They like to build their mud tunnels around the cylinders next to the engine case. This happy dauber dwelling location, however, can cause inadequate air flow cooling and result in a "hot spot" on your engine.

The C-152 engine cowling access door is small and, in my opinion, inadequate to make a complete visual inspection of the entire engine compartment. However, no matter which type of trainer aircraft you are pre-flighting, it is time to check the engine oil. Both the stock C-152 and Tomahawk models are equipped with variants of a Lycoming O-235 engine. The maximum amount of oil allowed in this engine is 6 quarts and the minimum is 4 quarts. It is a necessity for all reciprocating engines to have a crankcase breather tube so that amassed gases can escape. It is a quirk of most aircraft engines that the first quart of oil is blown out of this breather tube. Because of this, most operators usually allow the quantity of oil in these engines to remain one quart low. For the Lycoming O-235 engine, an accepted practice is to run the engine with 5 quarts, adding a quart when the quantity is reduced to 4 quarts. The C-152 checklist says to fill the engine to 6 quarts for extended flight. This being your flight plan, I would recommend doing so.

While we are on the subject of engine oil, do you know the nine things that engine oil does for an aircraft? The obvious benefit is providing lubrication. Did you know that oil also cushions, cleans, seals, cools, prevents corrosion, neutralizes acidity, affects weight and balance, and can be routed to do work for the pilot (such as with a constant speed propeller)? It does, but because of all these duties, crankcase oil will quickly become saturated with contaminants. It is recommended that engine oil be changed every 50 hours.

Some have suggested that in order to get an accurate oil level reading you must pull out the dipstick, wipe it off, re-insert it, screw it down, unscrew it, and then pull it back out to be read. I have found this to be true only after adding a quart. Try it for yourself, and use whatever technique you are comfortable with. There is one thing, however, that is a real pet peeve of mine. Do not over-tighten the dipstick when you screw it down. Notice that it is equipped with a rubber "O" ring that will seal with heat expansion. All that is necessary to properly seal this "O" ring is to lightly snug the threads. It is frustrating and sometimes knuckle-busting to attempt an oil check and be unable to do so because someone has over-tightened the dipstick.

If you are going to fly a C-152, now is the time to drain the fuel strainer. The strainer's job is to catch debris and/or water before it reaches the carburetor. Anything trapped here can be removed from the aircraft during preflight. The checklist says to pull the fuel strainer drain knob for about 4 seconds to clear the strainer and then check that the strainer drain is closed. What this means is that the strainer drain knob may not close completely from its spring loaded device and some fuel may continue to leak. Simply push on it to ensure that it is closed all the way.

I strongly suggest for two reasons that you catch the fuel sample in your sampler cup rather than let the fuel drain onto the ramp. First, you can examine the sample visually for any problems and then discard it away from the aircraft. Second, notice that the exhaust stack is right over the place where a fuel puddle would be. Now, stop and think that hot gases and quite possibly sparks come out of this stack during start-up. Allowing a fuel puddle to form increases the risk of fire.

If the engine is not already hot from a previous flight, grasp the exhaust stack and check it for tightness. The stack can loosen on some models.

Before you close the access or cowling door, inspect the wiring to the battery, solenoid, voltage regulator, alternator, right magneto, and spark plug wires. This check is easier on the Tomahawk than the C-152.

Photo 3-16
Latch the Tomahawk cowling securely

Look for any loose, broken or corroded connections, or frayed wires.

Make sure that when you close the access door on the C-152, it latches securely. On the Tomahawk, ensure that the locking latch goes under the tab before turning the locking wing nut (see Photo 3-16).

The obvious things to check at the nose of the airplane are the propeller and spinner. Gently grasp the spinner and try to move it up, down, and sideways. There should be no play or give whatsoever. If the spinner moves at all, it could end up coming over the cowling and in your face. If it wobbles because of loose screws, this is when having that type of fuel sampler with the screwdriver blades helps. Any loose spinner screws can be tightened easily without wasting time looking for a separate screwdriver and can be done by the pilot rather than seeking a mechanic.

Photo 3-17
Be sure to check the back of prop for damage

The traditional thing to do with the propeller is to run your hand along its leading edge to feel for any nicks, dings, or gouges. Check the back side of the prop blades (the ones that face the cockpit) because these faces of the prop are exposed to the ground and will likely be the ones pitted from foreign object damage (FOD) (see Photo 3-17). It has surprised me to find that some pilots do not know why they are performing this check and continue on to fly the airplane without understanding the ramifications of something amiss with the prop. Stop and think for a moment that the prop will be spinning in the neighborhood of 2300 to 2700 RPM. Along with this high speed spinning (the prop tips being barely less than sonic) are associated forces and stresses. These include centrifugal, torque, thrust, bending, and twisting stresses.

These stresses are normally absorbed along the entire length of the prop. Where there is a nick, gouge, or large pit on the propeller, there will be a localized stress. This local stress can cause a crack to develop and/or migrate across the prop. In some instances, a prop section separation can occur, leading to serious imbalance and quite possibly a compromise of flight safety when the engine is jerked off its mounting frame or firewall. So for those of you who did not understand why you check the prop, I hope that you get the picture now.

All that usually is necessary to fix a prop is to have an A&P mechanic inspect and, if necessary, file the nick down smooth (see Photo 3-18). This removes the stress and diminishes the risks. Do not file the prop yourself. An aircraft owner/operator, who is not an A&P, is not approved by the FAA to perform this maintenance. An aircraft that is operated out of a grass, dirt, or gravel strip is especially susceptible to prop FOD. One operated out of a paved strip is not entirely immune from damage, either.

Photo 3-18
Only A&P mechanics should file props

Look into the engine cooling air intakes for obstructions. Reach your hand behind the spinner and check the alternator belt. Just like your car alternator belt, it should have no less than 1/2 inch and no more than 1 inch slack when moderate pressure is applied. Inspect the landing light. You need to ensure that the landing light lens is not cracked or broken from rocks or debris picked up by the prop. If it is damaged, there may be glass shards present which, if left on the ramp or runway, can cut a tire. I prefer to have this light working, which I will discuss later, but there is no requirement that it actually operate to be legal for flight.

Moving down the nose, the next thing to check is the carburetor air inlet and filter. In the case of the Tomahawk, this air inlet makes a cozy home for birds (see Photo 3-19).

One time I had a late afternoon flight with a student in the Tomahawk and found signs of nest building activity. I used the opportunity as a learning experience for the student and, after cleaning the intake thoroughly, we flew. The next morning I had another instructional flight in the same aircraft. Pre-flight revealed a complete nest with total air intake blockage. I learned that overnight, a mama bird could be very busy indeed.

The air filter itself is an oil-soaked filter. It needs to be slightly oily (not dripping oily) to work properly. By its nature, though, it is a dust and dirt magnet. Check it for excessive dust or dirt air flow restrictions as well as for other wind-blown foreign matter.

It is wise to inspect the nose wheel tire and oleo strut for proper inflation and wear as discussed previously. All recommended tire pressures can be found in the "Handling, Service and Maintenance" section of the POH. Proper inflation of the nose gear oleo strut should leave about 3 to 4 inches of the piston exposed. One way to quickly estimate this inflation is to hold up three or four fingers, depending on the size of your hand, to the piston (see Photo 3-20). As mentioned before, wiping the oleo strut piston clean will help to extend the life of the seals. Remove the nose wheel chock or nose tie-down. The Tomahawk checklist recommends that you check and clean the windshield. This is a good policy for any aircraft at any time.

Photo 3-19
The front of a Tomahawk

Photo 3-20
Checking a C-152 nose gear oleo
strut for proper inflation

A quick note here about caution and safety around the propeller area. Your cautionary instincts should tingle when you are near this device. You must be ready and able to vacate the area in an instant. Even when you are bending down to inspect the landing light, nose gear, air filter, etc., it is a good practice to keep your hand on one of the blades so that when you stand up you will be less likely to hit your head. Reducing your vulnerability is the key, whatever your routine or practice.

Now we move to the forward cowling section on the left side of the aircraft. The first item found here on the C-152 is another small rectangular flap. This flap is opened by way of a screw that should be seen clearly. Here is another place where the fuel sampler with the screwdriver blades will be helpful. If a screwdriver is not available, a dime usually will work. Behind this flap is the connection for an external power unit (EPU) hook-up. In short, it is the place to hook up a battery for a jump-start. The battery for a C-152 lies directly beneath that small access door where you checked the oil. It is not hard to imagine that hooking up jumper cables through that access door would be next to impossible. So Cessna provided another way to link to the battery. The usual battery voltage on this aircraft is 24 volts, but check the POH for your aircraft to be sure of proper voltage and procedures before connecting any EPU or auxiliary power unit (APU).

At this station on the C-152, you will find its one and only static port. The port hole itself is elevated from the fuselage by a metal disc roughly the diameter of a Kennedy half dollar. The only functional explanation I can conceive for putting the static hole on a plateau such as this is to help disrupt the air flowing past it to achieve the desired static air status. Static pressure on any aircraft is routed through a closed plumbing system to the altimeter, vertical speed indicator, and airspeed indicator. By itself, this sampled static air gives us the aircraft displayed altitude (when properly set for the existing barometric pressure) and associated changes in actual altitude, which is shown on the altimeter.

Actual altitude can be expressed in two ways, *True* altitude and *Absolute* altitude. True altitude is the aircraft's height above mean sea level, and Absolute altitude is its height above the terrain. The memory aid I use to remember these terms is that the "T" in **T**rue has a flat top, representing a flat sea, and the "A" in **A**bsolute is peaked like the mountains on land. True altitude is read directly on an altimeter (indicated altitude), while Absolute altitude will require a computation of subtracting the True altitude of the terrain (or obstruction) from the indicated altitude of the altimeter.

Static air by itself also gives us our vertical speed in the atmosphere or, otherwise stated, our rate of climb or descent, which is displayed on the vertical speed indicator. Static air pressure routed to the airspeed indicator is registered with ram air pressure from the pitot tube to measure dynamic presssure which ends up as an indication of airspeed (See Figure 3-21).

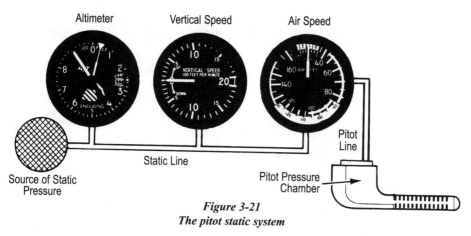

Figure 3-21
The pitot static system

Just above the static port, you will find the small air scoop that was mentioned before as a mirror twin to one on the right side. Also found at the leading edge of the wing root on the C-152 is the left fresh air and ventilation intake. Inspect these as carefully as you did the previous ones.

Now get up and visually check the C-152's left wing fuel tank for quantity. It is not necessary for this tank to have a vented fuel cap. It has its own separate fuel vent under the wing, which we will check later. If it does have a vented fuel cap like the one on the right wing, there is nothing wrong. This would simply be redundant venting for this fuel tank.

On the Tomahawk, raise the left side cowl and visually inspect the engine for bird and bee nests. Then check the left magneto and all electrical wiring for breaks and frays. On the firewall on the left side you will find the brake fluid reservoir. Check its level and take time to fill it to the proper quantity if it is low. Be suspicious of a larger problem if it is empty and/or low on a regular basis. If this is the case, investigate this problem with an A&P mechanic before flying the airplane.

Photo 3-22
The Tomahawk's fuel strainer housing

Below the brake fluid reservoir on the firewall of the Tomahawk is the fuel pump. Check all fuel fittings for leaks. Some of the fuel fittings which cannot be checked easily are those to the fuel strainer. The fuel strainer on the Tomahawk is housed in an enclosure inside the lower section of the left side of the cowling (see Photo 3-22).

Before closing the Tomahawk cowling access door, ask yourself these questions: Did everything look right? Am I compromising anything? Does everything feel right?

Learn to trust your feelings. Your subconscious may be telling you something important. Some people might consider this inclination of trusting your feelings as an allusion to Karma. But, after all, Karma is the doctrine of inevitable consequence. Remember, this is a machine and machines can break or wear out. So whatever your practice or doctrine, do not hurry. Learn to be methodical, careful, even suspicious when necessary.

Photo 3-23
The Tomahawk's fuel strainer drain

Just below the left cowling access door on the Tomahawk is a small NASA/NACA scoop, and below it is the fuel strainer drain

(See Photo 3-23). Drain some fuel and check it as before. This small air scoop is an interesting one which many people do not notice. One day a student asked me what this scoop was for, and I honestly told him that I did not know. When I could not find an explanation of its function in the POH, I embarked on my usual search for an answer. After asking a number of qualified people for an explanation of its function, I finally got a satisfactory explanation from an A&P mechanic fresh out of school. He said that fuel which is routed through a sharp bend in tubing will tend to heat up. He further surmised that the scoop was there to air-cool the fuel routing through this enclosure, thus preventing a possible vapor lock. Sure enough, when I finally had the

Photo 3-24
The sharp bend in the fuel plumbing inside the Tomahawk's fuel strainer housing

opportunity to see the plumbing inside this housing, I found that it does indeed make a sharp bend (see Photos 3-22 and 3-24).

Above and behind this fuel strainer area is the larger NASA/NACA air scoop, twin to the one on the right side. This air inlet routes ventilating ram air to the left seat in the cockpit. Check it as you did the others. Now we are ready to move down the left wing which is much like the right wing except for some small but important items.

Inspect the tire, brakes and oleo strut, if equipped, as you did on the right side. On the C-152, hanging below the wing, is a forward facing, bullet-shaped probe with a hole in the forward end. This is the pitot tube. The pitot tube samples ram air pressure as the aircraft moves forward and routes this ram air through a closed plumbing system to the airspeed indicator. As explained earlier, this ram air pressure is taken in differential with the static air pressure to register a pressure total or P-total. In engineering shorthand, this P-total is usually referred to as pitot. For aircraft piloting purposes, we know and use this result as the indicated airspeed of the airplane (see Figure 3-21).

Never poke a wire or stick into the pitot tube to try to clear an obstruction. Chances are that you will push the obstruction farther into the plumbing and lodge it tighter. Likewise, do not blow into the pitot tube. This may damage a pressure sensitive diaphragm in the airspeed gauge. If pitot tube blockage is suspected, the plumbing must be disconnected at a junction. Then compressed air should be blown toward the pitot tube from the airspeed indicator end to force the obstructing mass out through the inlet hole. Here again, this procedure should be done only by an authorized A&P mechanic.

Be sure to check the pitot tube drain hole. On most Cessna aircraft with the above mentioned type of pitot tube, the drain hole will be at the lower rear corner of the pitot pedestal, where it upturns into the wing. The drain hole allows precipitation or condensation to vacate the system before it gets into the business part of the plumbing.

Photo 3-25
C-152 left wing fuel tank vent tube

Continuing with the C-152, untie the left wing, and then look behind the left wing strut where there is a tube coming down out of the wing and making a turn, with the open end facing forward. This tube is the left wing fuel tank air vent tube. Check it for any of the previously described obstructions (see Photo 3-25).

At this point, you may be asking yourself why Cessna put a fuel tank vent tube on this wing and did not put an air vent on the right wing for the other tank. As I understand it, the left wing fuel tank air vent originally was to serve as air provider for both tanks, and there was a connecting vent tube running overhead the cockpit between both tanks. When the engineers designed this aircraft on paper, all the data and numbers indicated that this was a sufficient air supply to vent both tanks, and it probably was. Problems arose in the real world, however, as they usually do. When the aircraft was flown, this airflow plumbing arrangement did allow air to properly vent both tanks when the wings stayed level. But, when unusual things occurred such as turbulence, crosswinds, or slips to land, fuel was sloshed into the cross connecting tube. A combination of surface tension and capillary action held some fuel in this connecting tube, which in turn created an airflow blockage. The result was that all the fuel could be used from the left tank, but a vacuum created in the right tank prevented all of its fuel from flowing to the engine.

Obviously, any planned maximum range cross-country flight with this situation would be unexpectedly cut in half, not to mention the imbalance problems which would occur. In having to come up with a solution to this "faux pas," the engineers had a choice. They could redesign the fuel venting airflow system by adding a separate vent to the right wing. This fix, however, would have resulted in considerable and expensive production line and assembly retooling as well as additional FAA type testing. Alternately, the engineers could vent the right fuel tank with a simple AD requiring a vented fuel cap on the right wing. I believe you already know that they chose the simple (and less expensive) fix.

But let us give credit where credit is due. The engineers were thinking to a certain degree when they placed the aforementioned left wing fuel tank vent directly behind the left strut. This strut runs interference for the vent and helps prevent bugs, dirt, and water from entering and thereby becoming an obstruction to airflow when the aircraft is moving forward. This placement also helps avoid pressurizing the fuel tank with ram air. Hats off to the engineers.

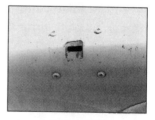

Photo 3-26
C-152 stall warning air vent

The last item to check on the C-152 wing that is different from the right wing is the stall warning horn air outlet. This outlet is seen as a small horizontal slit opening directly in front of the fuel vent on the leading edge of the left wing. The POH refers to this as a pneumatic-type system. More simply stated, it is operated by air pressure. More accurately stated, it works on low air pressure (see Photo 3-26).

Inside the slit is a reed-type device that operates on much the same principle as a clarinet or saxophone in that it vibrates when air is forced over it. In this case, however, the vibration and associated horn noise is produced from suction rather than from a blowing action. As the aircraft wing angle of attack increases, normal low pressure on top of the wing migrates forward to the leading edge of the wing, drawing air over the reed. The audible portion of the horn itself is located near the upper left corner of the windshield in the cockpit.

To check the stall warning horn, follow the POH's recommendation to place a handkerchief over the vent before you suck some air through it. This is a courtesy for the person who will check it on a subsequent flight, and it prevents you from sucking unpleasant bug parts into your throat.

To finish the C-152 pre-flight inspection, move around the left wing in the same manner that you did the right wing, checking the fairings, aileron, flap, hinges, and swivel joints. Do not forget to check a fuel sample from the left wing fuel sump at the rear portion of the wing root. In addition, there is one more fuel drain check that must be performed on some C-152s. An extra fuel drain may be found on the belly of the plane directly beneath the fuel shutoff valve. This fuel drain, when present, is positioned at the lowest point in the C-152's fuel system. This modification, added at an owner's discretion, must be logged in the aircraft's permanent airframe logbook at the time of installation. Also, it should be added to that aircraft's pre-flight checklist. However, in the event that this modification has not been included on the checklist, it is important that, prior to getting into the aircraft, the pilot visually inspect for the presence of this belly drain, especially on rental aircraft, and then drain some fuel from it to check for any water or contaminants.

Going back now to finish up the Tomahawk pre-flight, inspect the fuel vent, tire, brakes, and take a fuel sample as before. Untie the left wing and visually inspect the fuel quantity, remembering what the left gauge read initially in order to correlate its accuracy. The pitot tube on the Tomahawk is not a tube at all. It is referred to as a pitot mast. Ram air is accepted into the forward facing hole of the mast, while any precipitation is allowed to vacate through a downward fac-

Photo 3-27
Tomahawk pitot mast

ing drain hole (see Photo 3-27).

The stall warning horn on this aircraft is an electrically operated device, actuated by a mechanical tab on the leading edge of the left wing which closes an electrical switch when the wing is at a high angle of attack. When the wing is moving through the air at a low angle of attack, air pressure holds the tab in the down position. As the wing is moved to a higher angle of attack, air gets under the tab, lifting it up to close the switch. To audibly check the horn and tab for operation, the aircraft master switch must be turned on. Some pilots prefer to check the Tomahawk's stall warning horn as part of the "inside the cockpit" checklist while the master is already on. The practice of turning the master switch on at this point to check the horn should be followed with the practice of turning it back off again when the check is complete. Turning the master switch off to prevent battery drain is especially crucial in cold weather.

One word here about stall warning horns in general: they are warning devices only and should not be taken as an absolute. Ideally, the stall warning horn should be adjusted by an A&P mechanic to operate 5 to 10 knots above a stall in all phases of high angle-of-attack flight. I have personally experienced stall warning devices which sounded off at much lower tolerances than that. In fact, on several occasions in different aircraft, the horn was first heard as the stall was occurring and on one flight, the horn was heard only after the stall was felt and the recovery was initiated. So my advice is to never bet your life on a cheap warning device. Trust your life to good primary and recurrent training.

Finish the Tomahawk pre-flight with inspections of the fairings, aileron, flap, hinges, and swivel joints. With both the C-152 and the Tomahawk, you should be standing in the same spot where you initially started the pre-flight check. A habit that I have adopted is bending down and scanning under the airplane as a last double-check for any missed wheel chocks or untied tie down ropes. Quick, easy, and reasonable double-checks like this should become part of a pilot's normal procedure. If discrepancies or non-airworthiness items are found or even gut-felt, more thorough double-checks should be a pilot's normal way of life.

While you will find that there is variation in manufacturers' recommended procedures for various aircraft, such as the pre-flight, instructing you to proceed in one direction or another or to check different items due to differing systems, I strongly recommend that you follow the manufacturer's approved POH as much as possible. Remember, though, that the POH normally does not cover the minute details of pre-flight inspection as thoroughly as I have discussed them here. While it really does not matter which direction you proceed around your airplane, it does matter that all stations are visited and all items of the pre-flight process are thoroughly covered and checked to total satisfaction by the PIC of the planned flight.

"Complacency is one of the major causes of accidents. No matter how well things are going, something can always go wrong."

—Art Scholl, former Hollywood Pilot and Airshow Performer

CHAPTER 4

Winter Pre-flight

"A pilot has to believe in his equipment."

—Chuck Yeager, former Test Pilot

Although flying in cold temperatures has its rewards from a performance point of view, the pre-flight preparation in cold temperatures is much more important from a safety perspective. It is unfortunate that most pilots abbreviate this activity at a time when a detailed pre-flight is needed more than any other time. I know that it is tempting to take short cuts when you are freezing, but I find it best to think of winter flying as an advanced form of flying.

Abbreviating the pre-flight inspection can be a pitfall anytime of the year. Missed checklist items can result in an unplanned landing. The difference is that in the winter time, even if you have the good fortune to survive a forced landing, the cold can kill you and your passengers within hours. Do not let yourself fall into the quick pre-flight trap. Specially discipline yourself in the winter to check all the normal checklist items while giving consideration to the extra things that are discussed in this chapter.

The issues I will cover, however, do not necessarily constitute the definitive list of items on winter pre-flight activities for all aircraft. They are simply a place to start. Once again, I emphasize that my assessment of this topic is designed to make you think about the various aspects of winter pre-flight and the preparation of the aircraft that you fly during the cold months of the year.

The time to begin thinking about winter flying is before cold temperatures ever set in. Look in your manufacturer's POH and follow the recommended procedures for winterizing the aircraft. You can do some of this work yourself, but refer to the Federal Aviation Regulations (FAR) Part 43 for the specifics of what you can and cannot do. I recommend having a certified A&P mechanic carry out the required work. They do this work all the time, and their trained eyes may find other problems that might otherwise go unnoticed. Some of the specific items that I recommend for winterization are:

✈ install the manufacturer's recommended baffle plates to restrict cold airflow over the engine so that oil temperatures stay in the green

✈ check the heating and defrosting system for carbon monoxide (CO) leaks and carry an inexpensive CO detector (see Photo 4-1)

✈ check the engine breather system as well as all hoses, clamps, fittings, flexible tubing, and seals for deterioration

✈ check the battery for charge and capacity; cold temperatures rob a battery of its capacity

Photo 4-1
CO detector attached to instrument panel

✈ purchase or fabricate a pitot tube cover and engine inlet plugs to keep birds, insects and snow out (see Photo 4-2)

✈ remove the wheel pants where slush can accumulate and freeze

✈ consider getting wing covers if you park outside and need to keep wings free of snow and ice

✈ check the emergency locator transmitter (ELT) battery and

Photo 4-2
Engine inlet plugs can be made of scrap foam.
A pitot cover can be made from a tennis ball.

know how to manually activate the ELT

✈ prepare a simple survival kit and keep it in your flight bag. Some items to consider including are:

1. aspirin, band-aids, matches and candle
2. high energy foods such as granola bars and chocolate
3. a lightweight solar blanket
4. a small knife and strong line such as parachute cord
5. signaling devices such as a whistle and mirror
6. a quart of water in a plastic canteen
7. a booklet on shelters and survival
   (Note: Store the aspirin, band-aids, food, and matches in a strong zip lock plastic bag, taking care to purge all air.)

The planning continues with a good weather briefing from an FSS. Be sure that you understand the weather forecast report for your destination and at your home airport if you plan to return. Filing a flight plan (Figure 4-3), using Air Traffic Control (ATC) flight following and telling someone when you plan to return are all forms of cheap life insurance. You should use one or all of these methods at any time of the year but in winter they may be your ticket to survival.

In your pre-flight planning, locate any alternate fields that could be quickly and easily utilized. Allow greater than the legal minimums for fuel reserves. Stay night current, because days are shorter in winter. Finally, adopt the mindset that you will use any options and alternatives available, even if that means making a 180 degree turn and returning home.

Before you go out to the aircraft to conduct the winter pre-flight, it is important that you dress properly. The more comfortable you are, the less likely it is that you will rush through the procedures. Dress in layers so that you can shed clothing later when the aircraft heater is working. Wear a hat or toboggan. At 40° Fahrenheit and lower, fifty percent or more of body heat is lost through the head. Some other items to wear are a scarf, fur lined gloves, parka with a hood, and long john underwear.

Boots with good tread are essential. No one can perform an aircraft pre-flight properly while slipping and sliding on an icy ramp. Imagine if you had to use a tow bar to pull the aircraft into position for starting or

U.S. DEPARTMENT OF TRANSPORTATION FEDERAL AVIATION ADMINISTRATION					TIME STARTED	SPECIALIST

FLIGHT PLAN

Pilot Briefing/Flight Plans — 1-800-992-7433

1. TYPE	2. AIRCRAFT IDENTIFICATION	3. AIRCRAFT TYPE/ SPECIAL EQUIPMENT	4. TRUE AIRSPEED	5. DEPARTURE POINT	6. DEPARTURE TIME		7. CRUSING ALTITUDE
VFR					PROPOSED (Z)	ACTUAL (Z)	
IFR							
DVFR			KTS				

8. ROUTE OF FLIGHT

9. DESTINATION (Name of airport and city)	10. EST. TIME ENROUTE		11. REMARKS
	HOURS	MINUTES	

12. FUEL ON BOARD		13. ALTERNATE AIRPORT(S)	14. PILOT'S NAME, ADDRESS & TELEPHONE NUMBER & AIRCRAFT HOME BASE	15. NUMBER ABOARD
HOURS	MINUTES			
			17. DESTINATION CONTACT/TELEPHONE (OPTIONAL)	

16. COLOR OF AIRCRAFT	CIVIL AIRCRAFT PILOTS. FAR Part 91 requires you file an IFR flight plan to operate under instrument flight rules in controlled airspace. Failure to file could result in a civil penalty not to exceed $1,000 for each violation (Section 901 of the Federal Aviation Act of 1958, as amended). Filing of a VFR flight plan is recommended as a good operating practice. See also Part 99 for requirements concerning DVFR flight plans.

FAA Form 7233-1 (8-82) **CLOSE VFR FLIGHT PLAN WITH _____ FSS ON ARRIVAL**

SPECIAL EQUIPMENT SUFFIX

/X — No Transponder
/T — Transponder with no altitude encoding capability
/U — Transponder with altitude encoding capability
/D — DME, No Transponder
/B — DME, Transponder with no altitude coding capability
/A — DME, Transponder with altitude encoding capability
/G — GPS, Global Positioning System

/M — TACAN Only, but no transponder
/N — TACAN Only and Transponder, but with no altitude encoding capability
/P — TACAN Only and transponder with altitude encoding capability
/C — RNAV, Transponder with no altitude encoding capability
/R — RNAV, Transponder with altitude encoding capability
/W — RNAV, No Transponder

SPECIAL EQUIPMENT PREFIX

T/ — TCAS, Equipped aircraft

Figure 4-3
The flight plan format with all current special equipment suffixes and prefixes

into a heated hangar for the pre-flight. Without good traction, you could provide the coffee drinkers in the FBO with quite a show. In the process you could injure yourself on the spinner, propeller, or you could slide under a moving nose wheel.

The absolute best way to keep your aircraft free of snow and ice is to keep it in a hangar. The only danger with a heated hangar comes from the practice of topping off the tanks before putting the aircraft inside. Most people do top off their tanks in the winter, because this reduces space in the tank for condensation and thereby prevents water from being introduced into the tanks. Cold fuel will expand when warmed, however, and if the aircraft is fueled while cold and then is put into a heated hangar, the fuel will expand and could be forced out the overflow or air vents. This produces a fire hazard. The solution to this problem is to refuel to within a half inch of the top of the tanks to allow room for the fuel to expand but not overflow.

Let us assume that the aircraft you will fly sits on the ramp exposed to the elements. A heavy snow will probably find you at home where it is nice and warm. If you own an aircraft, you should consider dressing warmly and making a sojourn to the airport to remove the thick snowfall from the wings and horizontal stabilizer. Heavy stress loads from this added weight are unnatural to the aircraft and may cause structural damage.

If you plan to fly, one of your options for snow, ice, or frost removal is moving the aircraft into a heated hangar for the pre-flight inspection. While you are performing your other duties, the added warmth will help melt accumulations from the aircraft surfaces. Be sure, however, to completely dry the skin and control surfaces before you move the aircraft back into cold air, otherwise any remaining water will re-freeze and either reduce lift, lock up the control surfaces, or both, defeating the purpose of using the hangar.

Never attempt to take an aircraft off the ground with frost, snow, or ice on the airframe. These wintry encrustations can reduce lift by 15% or more. Accident files are littered with too many reports involving all types of planes, from the smallest general aviation aircraft to large airliners, describing unnecessary accidents related to icy conditions. Think of ice, snow, and frost as foreign substances to the aircraft and not natural for flight. They can be effectively removed, however, even when the airplane sits on the ramp continuously. Some of the things to consider before attempting foreign substance removal are:

✈ what foreign substance is to be removed—frost, snow or ice?

✈ which tools or equipment may be required (different foreign substances mandate different techniques)?

✈ is it a high wing aircraft or low wing aircraft (a high wing or a T-tail will require a ladder for success)?

The specific kind of foreign substance that coats your airplane at the time will determine what action you should take to remove it. Frost

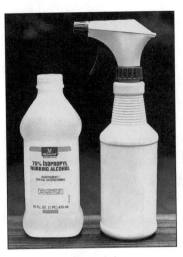

Photo 4-4
A simple removal system
for frost

is easiest to remove. A simple household spray bottle filled with inexpensive 70% isopropyl alcohol will cause frost to disappear magically when sprayed on the wings, fuselage and empennage (See Photo 4-4). Do not use alcohol on a plastic windshield, however. It will cause the plastic to turn milky over time and will require its eventual replacement. Although alcohol works well on frost, it will not work on snow or ice. It is thought by some that frost can be "polished" smooth and flight can commence immediately. While this may help a little, nothing is better than total removal of any foreign substance from the aircraft.

When snow is the surface invader, it sometimes can be wiped away with thick work gloves, a stiff brush, or a broom. Be careful, though, because a layer of ice can be hiding on the aircraft surface beneath the snow. It is acceptable to concentrate your efforts on the flight and control surfaces, but do not neglect cleaning the entire exterior surface of the aircraft. Snow and/or ice on the fuselage can affect weight and balance. Remove snow and snow melt from all air intakes and heater ducts. Some aviators do not worry about removing snow from the engine cooling inlets because they believe the rising engine temperature will melt it. This kind of thinking may result in a cracked engine block as the engine temperature rises too fast from an initially restricted airflow.

By far the hardest culprit to remove is ice. If a very thin layer is present, it may be scraped away with a non-abrasive rubber tipped squeegee. A harder or thicker layer may be coaxed away with a hard plastic, automotive type scraper, but this type of scraper never should be used on the windshield because it will scratch it. Whatever technique you use to remove an ice layer, do not use your hand, arm, or anything else to bang on the wings, fuselage or control surfaces in an attempt to crack it off. Fiberglass wing tips and fairings are brittle and fragile from the cold, and they crack very easily from even a slight impact. Light airplanes are not built to be banged on, especially when they are stiff from winter's cold.

Another thing that you absolutely should not do is use any kind of water to remove snow, ice or frost. Do not use a hose with running water, and do not use a bucket or pitcher with hot or warm water. The water will run into joints, hinges, control rods, and other crevices and freeze, causing dimished or loss of control. I do not know a single pilot who knowingly wants to hurl a hunk of metal through the air unless he or she is in total control. Remember that if it is cold enough to freeze, it is cold enough to re-freeze. Keep in mind also that it will be even colder at cruise altitude.

As you continue through the pre-flight, check the freedom of movement of the control surfaces from outside the aircraft as well as from the inside. Check the inside of the propeller spinner as best you can for any ice or re-frozen snow melt. The presence of either of these substances can cause a serious imbalance when the engine is started and could result in vibration damage. One way to help avoid this scenario is to leave one prop blade sticking straight down so that any precipitation will run out of the spinner instead of accumulating there.

The pitot and static inlets can become blocked with snow and ice in the winter. This obstruction usually can be wiped away from the static port. For the pitot tube, a short blast of pitot heat will melt even the hardest ice. In addition, these areas also can become havens for small insects escaping the cold, causing internal blockage. As stated previously, never blow into the pitot or static openings. If a blockage is suspected that cannot be cleared externally, disconnect the plumbing and use compressed air to blow it from the cockpit outward. Again, have your A&P mechanic do this for you.

Another insidious possibility is frozen precipitation clogging the fuselage belly drain holes. When this happens, the fuselage will be unable to drain precipitation or condensation, and a build up of frozen matter will result. If the accumulation is great, it can throw the CG rearward depending, of course, on the arm. On the Piper PA-28 series of airplanes (Cherokees, Warriors, etc.), the stabilator has drain holes on the underside of its leading edge. It does not take much freezing rain weight

Photo 4-5
Drain holes on a PA-28 stabilator

accumulation here at the most rearward arm to produce a significant CG change and/or adverse handling characteristics (see Photo 4-5). Any subsequent flight would most likely be unsafe with this situation.

I continue to be amazed and saddened by the number of accidents caused every winter by water in an aircraft's fuel system. We may be lulled into complacency because of an assumption that gasoline and water do not mix. We are taught to believe this when the instructor first shows us a fuel sample where water was drained from the tanks. We saw how the water sank to the bottom, while the fuel stayed at the top with a distinct dividing line between the two. However, nothing could be farther from the truth. Water will indeed mix with fuel.

While it is true that fuel and water do not freely mix, fuel will absorb small but significant quantities of water, sometimes directly from the atmosphere. That water will mix into the fuel solution. The amount of water dissolved in fuel depends on two things: the temperature and the fuel constituents. The temperature factor should be easy to understand. Water will saturate fuel the same way that it does the atmosphere. When warmer, fuel will hold more water in solution. The opposite is true with colder temperatures. Note here that auto fuel will hold more water in suspension than aviation gas (avgas) on a given day.

In the atmosphere, we refer to the dew point as the temperature at which moisture will condense. In fuel, this condition is known as the saturation limit. If plunging temperatures cause the dissolved water concentration to exceed its saturation limit, the water is squeezed out of the fuel solution. Do not be fooled into thinking that this cannot amount to much water. A teaspoon of water in the form of ice crystals is enough to block a gascolator screen, causing the fuel pump to collapse the screen and thus preventing almost all fuel flow to the engine.

In order to understand fuel constituents, first you must know that fuels are not pure chemical compounds. They are a mix of numerous hydrocarbons blended together to achieve a performance specification.

This performance specification defines the fuel's use and suitability for certain engines. One way of boosting a fuel's octane rating is by adding aromatics. Automobile unleaded gasoline usually will contain around 25% aromatics. While the water solubility of various gasoline components varies with the blend, the aromatics that are added to the fuel can, because of their molecular structure, absorb six times as much water as the other hydrocarbons.

Some fuel refineries normally do not use aromatics in aviation fuel in order to minimize water solubility, but they are not prohibited by law from doing so. On the other hand, automobile gasoline is manufactured without aviation in mind and, during cold weather operations, should only be used in an aircraft with extreme caution. Many northern operators will refuel their aircraft through a chamois cloth to separate water from the fuel. While the chamois does work to a degree, its largest limitation is water saturation, after which point the chamois then becomes ineffective. I strongly recommend that when you drain fuel from the tank sumps in the winter, you drain more than the amount that you normally would during other times of the year. Even with this precaution, however, be aware that there may be some frozen water in the bottom of the tank which you cannot drain.

Some people advocate the use of alcohol to dry up water in the fuel. Alcohol has the unique property of mixing with both water and fuel. It will mix more readily with water than with gasoline, but it will dissolve in either. Alcohol dryers are best used within a few hours before a flight, because they will themselves absorb moisture from the atmosphere. Still there are some added precautions to take when using alcohol as a water dryer in fuel. Do not use methanol. It is corrosive. Do not use 70% isopropyl rubbing alcohol. It is already 30% water and therefore its absorption capability is nearly exhausted. It may add even more water to your fuel than is already there.

The best alcohol for drying out gasoline is 91% isopropyl (see Photo 4-6). It is relatively easy to find at the drugstore, is only pennies more than other alcohols,

Photo 4-6
Use 91% isopropyl as a fuel dryer

and is much more effective for fuel drying purposes. But do not use alcohol indiscriminately. Over time, it will cause deterioration of neoprene seals and other fuel system components.

Finding water in your fuel tanks on a regular basis may be reason for further investigation. It could be coming from a bad seal on a fuel cap, allowing rain to run in. It could be pumped in from leaking underground fuel tanks or be the result of condensation from the fuel truck itself. Talk with other pilots who use the same fuel you do and compare notes about similar problems. Try not to refuel during falling precipitation. The bottom line is to do everything possible to ensure that there is no water in your fuel system, frozen or otherwise.

Continuing the winter pre-flight procedures outside the aircraft, it is always a good idea to turn the propeller through a few revolutions to "break loose" or "limber" the oil in cold weather. Be careful, though, to stay clear of the propeller arc. Pre-heating the engine will aid in limbering the oil. Thick, molasses-like oil will not lubricate properly or quickly enough to prevent progressive damage to the engine. While this friction wear may not cause any immediate problems, it is cumulative, permanent and irreversible.

There is some controversy, however, as to whether the prop should be turned in a forward or backward direction. Some people say that the prop always should be turned forward, because turning it backward will damage the brushes in the alternator (or generator if your aircraft is so equipped). This is not true. It is true that the electrical contact brushes on an electrically heated propeller may be damaged by turning the prop backward, but on most light single engine aircraft this is not a concern. Even so, one part of almost any aircraft piston engine which could be damaged by turning a prop backward is the vacuum pump. The angle of the vanes in the vacuum pump and the plastic shaft make it susceptible to breakage. Would you want a weakened vacuum pump to fail when you need it most? VFR conditions might not be a cause for concern. Flying into an Instrument Flight Rules (IFR) condition, on the other hand, is a more serious situation. Consider the conditions into which you will be flying.

There is an advantage, however, to turning a prop backward. An aircraft engine will not start when turned backward. Let me explain. When a starter is engaged, it turns the crankshaft, which turns the gears, which then turn the magnetos, which develop the sparks that start the engine. The only problem is that the starter cannot turn the engine rapidly enough to turn the magnetos fast enough to create the spark that the engine needs to fire. The solution is a spring-loaded device inside the magneto called an impulse coupling. The impulse coupling will momentarily hold the magneto back from turning and then let it go with a quick impulse revolution, causing the hot spark that an engine needs to fire. Because the impulse coupling will engage only in the forward direction, the engine will start only when the prop is turned in the forward direction. Therefore, turning a propeller backward eliminates the possibility of an engine starting unexpectedly.

Another argument by those who advocate turning the prop in the forward direction is that "there is no fuel in the cylinders because the mixture was fully leaned to shut the engine down." With no fuel, there is no way that the engine can start. This is true if it is a carbureted engine, or if the last person to fly the plane was patient enough to wait until all of the fuel was burned before switching off the magneto switches, and if there is no broken magneto ground wire. But if there is a broken magneto ground wire, and if, for any reason, there is fuel in the cylinders, then the simple act of hand turning the prop in the forward rotation can result in anything from a broken arm to mortal injury should the engine fire even briefly. Ask your mechanic or instructor to explain this scenario to you further, if necessary. The bottom line is to reduce risk to life and limb. If you turn the prop backward, know the ramifications. If you turn a prop forward, do so with extreme caution and with the aircraft securely tied down and wheel chocks in place.

In the previous chapter, I mentioned the crankcase breather tube that allows accumulated gases to escape from the crankcase. In the winter, this breather tube can become clogged with frozen condensation. This type of clogging will lead to pressurization of the crankcase which can force engine oil overboard. Check the tube (by squeezing it or by using a dowel) up its length as far as you can reach to check for blockage, which may be located someplace other than at the tube end. It is

Photo 4-7
C-152 crankcase breather tube

extremely important to check this crankcase breather tube before a winter flight (see Photo 4-7).

If your aircraft has external baggage storage, keep in mind that all baggage stored there will be inaccessible during flight. One winter flight with an inoperative heater will convince you to layer clothes on your body and shed them as necessary as the heater takes effect. Be sure, however, to keep these clothes accessible in case the heater decides to divorce you. Another good idea is to keep a blanket or sleeping bag handy in the cockpit for yourself and any novice passengers. They will perceive you as a careful, well prepared pilot, and you could become a hero for the day. Cold passengers make it a miserable flight for everyone.

Before attempting to start your engine in cold weather conditions, refer to and follow the manufacturer's checklist and procedures. If the aircraft's battery is weak and an external power source is used to start the engine, be especially careful to read and follow the POH procedure for this operation. Some aircraft are supposed to have the master switch on while some should have it off during the process. Re-check the POH for the battery's voltage. Some aircraft have a 12 volt battery and some have a 24 volt battery. This is true in the case of our two example aircraft. The C-152 has a 24 volt battery while the Tomahawk has a 12 volt battery.

Winter flying can be the most beautiful and gratifying type of aviating that you will ever experience. On the other hand, it is different and can be very unforgiving. It is not necessarily more hazardous, though, if the pilot uses extra caution, exercises good judgment by thinking things through, and prepares and plans well.

Never be too proud to walk back to the FBO and get another cup of coffee. If you do not feel comfortable with the plane, with yourself, with your abilities, or with the weather, DO NOT GO. If you do go, it must be only after a complete and thorough pre-flight. Do not ever shrug and say you don't have the time for a pre-flight. No matter what season of the year, your time is never better spent than on a complete, careful, and thorough pre-flight. Remember: acting on the ground is preferable to reacting in the air.

"If you want to grow old as a pilot, you've got to know when to push it, and when to back off."

—Chuck Yeager, former Test Pilot

CHAPTER 5

Engine Starting and Taxiing

"Aviation in itself is not inherently dangerous. But to an even greater degree than the sea, it is terribly unforgiving of any carelessness, incapacity or neglect."

—André Priester, circa 1939

Once you have completed the pre-flight inspection and only after you are satisfied that the aircraft is reasonably safe to fly, it is time to get in the left front seat of the cockpit and get out the "Before Starting Engine" checklist. Always use a checklist no matter which aircraft you intend to fly. Just as with the pre-flight checklist, the "Before Starting Engine" checklist also differs somewhat between the C-152 and Toma-hawk. The Tomahawk checklist states up front that you should close and latch the cabin doors, while the C-152 checklist says nothing about clos-ing the doors. Does this mean that closing the cabin doors upon entering the C-152 is not relevant or important? Of course not.

It is natural for us to enter any vehicle equipped with doors and close them after entering. For the C-152, closing the door is all that is required to latch it, so the authors of that checklist probably assumed that it was not important to include it in the checklist. For the Tomahawk, however, latching the door after closing it is more involved. It therefore became necessary to include written commands in the checklist to ensure that this act was completed for both of the side door latches and the overhead latch (see Photo 5-1).

Another difference in the checklists of these two aircraft is related directly to the fuel delivery systems: high wing vs. low wing. Aside from these aspects, the majority of procedures used are the same or similar. However, some superficial differences still remain between the two aircraft. These lie mostly in the cockpit arrangement and/or placement of knobs, levers and switches. Nevertheless, the following procedures that I will outline can be used and related to any normally aspirated, direct drive, air cooled, horizontally opposed, carburetor equipped engine in nearly any aircraft.

Photo 5-1
The Tomahawk (pictured) has a more involved
door latch system than does the C-152

The next items on the two checklists are in agreement. Adjust your seats, seat belts and shoulder harnesses. I am surprised at the number of people who consistently gloss over this part of the checklist. They read "seat belts, shoulder harnesses" and even call this out when vocalizing through the checklist. Rather, here the checklist specifically addresses three items: seats, seat belts, shoulder harnesses. The first of these, and the often overlooked item, is the adjustment of the seat position. When adjusting the seat, not only should you get it comfortable for reaching the rudder pedals, but it is imperative that you ensure that the seat is locked in that position by attempting to move it fore and aft. If it is not securely locked into its track position, it can move later at an inopportune time. One such scenario might go something like this:

Imagine that you are speeding down the runway and have just rotated the nose for takeoff. As you sink slightly aft in the seat from the vertical acceleration, the seat shoots all the way back in a flash. Your automatic impulse makes you grip tighter onto whatever you can reach to stop the unexpected rearward slide. The only problem is that you are already holding onto something with both hands: the yoke and the throttle. As you grip tighter during the seat slide, you pull the throttle and the yoke all the way back. Now here you are, less than fifty feet off the ground, nose high, and with no power. This is an ugly picture which can be prevented by simply checking for seat and seat track security before you fly the aircraft. It does pay off to notice these details, as itemized, on the checklist.

Also in the "Before Starting Engine" section of the POH, you must check the circuit breakers, ensure the carburetor heat is in the _OFF_ position, and check that the fuel shutoff valve is in the _ON_ (horizontal) position in the C-152, or place the fuel selector on the Tomahawk to the desired or proper tank. Usually the proper tank on a two-tank, fuel selective system is the one that is most full. However, one instance where you might choose the less full tank is when there is already more weight on that side of the aircraft, such as when you fly with a large passenger. This is one of those situations where the PIC must make a discretionary judgment call. I suggest choosing on the side of better weight and balance initially and then possibly posting a peel-off note on the dash to remind you to check and/or switch the tanks in flight at a pre-determined time.

One other thing required by this section of the checklist is to set the parking brake in the Tomahawk and testing and setting the brakes in the C-152. Brakes work on the principle that a fluid will not compress. When the brake pedal is depressed, the pressure exerted on the brake fluid forces the brake pads toward the brake disc. Holding the pressure will hold the brakes. The brake fluid is housed within a closed system from the reservoir to the pads. If there were a leak in the system, the pads would not hold because constant pressure would be compromised.

Being a high wing plane, the C-152 offers one advantage in that the brake system can be checked for leaks by simply looking out the window for any fluid seeping out of any fittings when pressure is applied to the brake pedals. The pressure build-up in a brake line is such that if there is a leak in the system, you will see it. Many of us have heard of gremlins. These are imaginary creatures to whom unexplained mechanical problems are sometimes attributed. Indeed, I can attest that on at least one occasion, gremlins must have visited the C-152 in which I instruct.

Preparing for an early morning flight, a student and I reached this phase of the checklist. When the brake pedals were depressed, fluid squished out from the right gear brake line fitting. The aircraft, having been flown the day before, was grounded until fixed, and this student learned a lesson more valuable than any time in a logbook. On the Tomahawk, however, your should look for brake fluid leaks before boarding. The low wing aspect of this craft makes visual recognition of this potential problem nearly impossible once you are inside the cockpit.

The next section of the checklist deals with starting the engine. We will assume that the engine has not been started within the last few hours and that this is a cold start. While the factory-stock C-152 and factory-stock Tomahawk have the same basic Lycoming O-235 engine, the carburetor arrangement on these two aircraft is different. The C-152 carburetor has an accelerator pump, while the stock Tomahawk does not. The Tomahawk carburetor can be modified to incorporate an accelerator pump, but it did not come from the factory with one. While this difference will not affect the actual priming of either engine, it does indicate that different aircraft manufacturers' seemingly identical engines may actually be quite different. Therefore the checklists may differ in the order of their recommended starting tasks. For example, setting the mixture to rich is first on the C-152 checklist but fifth on the Tomahawk checklist. In addition, even though both checklists specify that the throttle be opened 1/2 inch, they do so at different places. I suggest that you follow the checklist order for the aircraft that you are manipulating at the time.

The primer is one item that I see repeatedly misused. Acting like a syringe, the primer pulls a vacuum in its chamber when it is pulled out. Do not pull the primer out and immediately push it back in. It takes a few seconds for the fuel to fill this vacuum, so be patient. If you want to see this principle in action, get a syringe (without the needle) and a cup of water. Place the tip of the syringe in the cup of water and pull the plunger to the back of the chamber. You will see that it does not fill immediately. The C-152 has a long draw primer, while the Tomahawk has one with a short draw. These two primers act exactly alike, and the only difference in the two is the fact that they have different manufacturers. (By the way, you should never prime with the throttle!)

When turning the primer, always turn it clockwise. If it is turned counter-clockwise and the shaft key gets momentarily caught in the locking collar key-way, the locking collar can loosen. A loose primer will make for an unhappy engine and may even cause engine failure. Also, after pushing the primer in for the last prime stroke, make sure that you rotate the primer shaft clockwise for approximately 180°. This can be done by placing your thumb on the primer, pointing down, and then turning the thumb up clockwise (see Photo 5-2).

Another simple means of accomplishing this 180° primer shaft turn is to look and see if the primer has writing on it. Find which letter is up and then turn clockwise until that letter is down. If you are one of

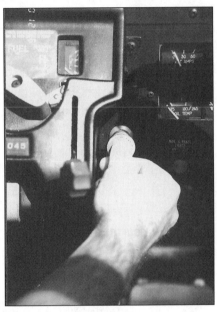

Photo 5-2
Be methodical when locking the primer

those folks who plays roulette with the primer by spinning it a few times, you may be surprised one day when it ends up barely locked and the start-up vibration moves the shaft key into alignment with the key-way. Be methodical in your actions and think about what you are doing. Your careful attention will pay off by reducing risk.

The C-152 checklist calls for you to "CLEAR" the propeller area, while the Tomahawk's does not. Not only is this modus operandi a time honored tradition for aviators, but it is an opportunity to bark out a warning that will be heard and recognized by anyone who may be walking up to your aircraft on your blind side. Do not overlook the importance of saying "CLEAR" or "CLEAR PROP" loudly and distinctly.

As a back-up to keep the aircraft from moving after starting, depress the brake pedals just in case the parking brake is weak and also have any passengers look around for people or obstacles. Turn the master switch on (for those flying the Tomahawk, turn on the fuel pump as well), place your right hand on the throttle, left hand on the key and engage the starter. Do not pump the throttle repeatedly at this point. To do so in the C-152 would actuate the accelerator pump in the carburetor, causing fuel to pool in the carburetor and the carburetor box. Should the engine backfire, a fire could result. In a stock Tomahawk, pumping the throttle has no effect since there is no accelerator pump. I will discuss the advantage of having an accelerator pump later in the chapter on imminent situations.

Once ignition has taken place, the left hand is finished with starting. Let go of the key and place the left hand on the yoke. The right hand gets busy immediately. Do not let the engine race at a high RPM. Immediately reduce power to achieve 800 to 1200 RPM in the Tomahawk and 1000 RPM or less in the C-152. A rate of 800 to 1000 RPM is more engine-friendly because of the initial lack of oil circulation. On a cold winter day, however, you may have to increase to 1200 RPM if you have a cold-natured engine. In every case, the engine oil pressure should move into the green arc on the gauge within 30 seconds of engine ignition. If it does not, shut the engine down immediately or the metal components will overheat and permanent engine damage will result.

It is now time to turn on the flashing beacon or strobe lights, as well as the radios and transponder. If your aircraft has both a red beacon and white strobe lights, I suggest turning on only the beacon. Strobes, where possible, should only be turned on just before departure because they are too distracting to other pilots in the vicinity. I will make one note here in reference to the rotating beacon. As you advance in the aviation ratings or attend some training facilities, you will learn to turn on the beacon before starting the engine. In turbine-powered aircraft, this is a requirement. For the purposes of this book, I am following the procedures per the current policy for trainer aircraft at the school where I instruct. I reserve the right to change any current practice should the need develop.

Now, in the C-152, it is time to raise the electric flaps. Recall that we left them down during the pre-flight inspection. Look back and watch them as they come up to see if one is lagging or dragging. As mentioned before, while you are in the air is no time to discover a split flap condition. Now that the engine is running, raising the flaps will not tax the battery needlessly.

In the Tomahawk, turn off the electric fuel pump and watch the fuel pressure gauge. A slight drop may occur, but not a large one. What you are doing is checking to see that the engine-driven fuel pump is working properly. The electric fuel pump was needed initially to get the fuel from the low wing up to the engine for starting. It will be required again in the Tomahawk for back-up on takeoff and once again for back-up during the landing phase.

The time has arrived for the first radio usage. I recommend, at the very least, using earplugs in an airplane. Hearing is too valuable a sense to be lost on a cumulative basis from the noisy drone in a cockpit. If available, headsets are a much better option. They not only protect from noise, they allow for a much clearer transmission and reception. If you are at a towered (controlled) field, you would listen to the Automated Terminal Information Service (ATIS). If you are at a non-towered field which has an AWOS installed, listen to it. This procedure is especially important for new pilots or students. It gets them in the habit of accessing available information, and it also decreases the initial anxiety of utilizing the radio. For the remainder of this discussion, I will assume that we are at an non-towered field unless stated otherwise.

Next it is time to call the FBO. I have repeatedly heard pilots ask only for a "radio check." This practice usually garners the response of "loud and clear" from the FBO, which confirms that the radio is working but leaves the pilot with little else in the way of information. Rather than teach my students to ask for a simple radio check, I have them ask for an "airport advisory." This request elicits a more informative response providing the active runway and any known traffic. At the same time, the student is assured, through natural assumption, that the radio is operationally "loud and clear."

A comment here upon hearing the words "no known traffic" or "no traffic" as an FBO response to an "airport advisory" query. These phrases do not guarantee that there is no traffic around. Rather, they mean that there is no KNOWN traffic. There may be a no radio (NORDO) aircraft entering the pattern, unknown to anyone. Alternately, there may be an aircraft with a radio that is not in use. Never relax and assume that there is no traffic around. You must always be vigilant, wary and wise. You can now announce to the airport traffic and indeed to the "no traffic" your intentions to taxi to the active end of the runway.

Taxiing is one of those activities that is difficult at first but becomes easier with practice. What most people do not understand is that very little power is necessary to taxi. An RPM setting of 800 to 1000 RPM is perfect for most taxi requirements. While it may require a little more RPM to initiate aircraft movement, many people err by leaving the power setting at the higher RPM. This accomplishes nothing more than premature brake wear, because the pilot must ride the brakes to keep from going too fast. A simple reduction of power to 800 RPM after initial aircraft movement will eliminate this need to ride the brakes. On the other hand, do not let the RPM fall below 800 for extended periods of time or the cylinders may load up with carbon. There is almost always a trade-off in aviation. A smart pilot will always use knowledge as the foundation for trade-off decisions.

The Tomahawk and the C-152 each will feel different when taxied, because each has different linkages from the rudder pedals to the nose wheel. The C-152 has a spring linkage which compresses when a rudder pedal is depressed. This system allows full rudder deflection without movement of the nose wheel. The Tomahawk, on the other hand, has a direct push/pull metal rod linkage. Anytime a rudder pedal is depressed, the rudder moves and the nose wheel moves, too. Unless the aircraft is moving, it will be difficult to depress the Tomahawk's rudder pedal because of the friction of the nose wheel on the pavement. For this reason, do not force the rudder pedal while the Tomahawk is at a standstill, or you may bend the linkage.

The very first thing that should be done after initial aircraft movement and power reduction is to check the brakes for effectiveness. If they are not effective, act immediately by pulling the mixture to stop the engine before you taxi into another parked aircraft. Once satisfied that the brake check is done, proceed to the runway. Again, do not ride the brakes. Use the rudder primarily to steer the airplane and only apply differential braking as necessary to help steer. While on the ramp, taxiing may involve negotiating through a maze of parked aircraft, so the taxi speed should be slow. Eight hundred RPM works fine. A rule of thumb when taxiing on the ramp is to taxi no faster than a person would walk. Another way to visualize this concept is to taxi at a speed at which you could react and stop the aircraft without damage to other aircraft. This is roughly a reaction distance of no greater than three to five feet.

Once you are clear of other aircraft and on the taxiway, you can go a little faster, but still no faster than a speed at which you could react and stop within a distance of eight to ten feet. This taxi speed can be achieved easily with a power setting of 800 to 1000 RPM also because you most likely will be taxiing with a tailwind to get to the end of the runway. If the tailwind is strong and the aircraft does move too fast, resist riding the brakes. Remember, constantly riding the brakes will overheat them and cause premature wear. It is better for brake life if, after the aircraft begins to accelerate slightly, the pilot actuates the brakes to slow the aircraft and then releases them to allow a cooling time between braking periods.

The aircraft always should be taxied with the nose wheel on the taxiway centerline. In almost all cases, this placement assures obstruction clearance for the aircraft. Looking at it from a legal point of view, if the pilot is on the taxiway centerline and hits an object, someone else may have to pay. If, however, the aircraft is off the centerline and does damage to property, the pilot may be held liable. I tell beginners to imagine the yellow taxiway stripe running right through the center of the cockpit directly between us. This gets their head up to look outside the aircraft and encourages them to see the "big picture." I heard an FAA inspector once say, "You can always tell a Captain by the way he taxis on the centerline." We should all strive to become more Captain-like.

If the airport at which you operate has no taxiway, then back-taxiing down the runway may be your only option. Back-taxiing is aircraft movement down an active runway in the opposite direction from that which landing traffic would use. Obviously you would not back-taxi if someone were landing, but back-taxiing down the centerline of the runway is important. This positioning serves to give other airborne aircraft their greatest opportunity to see you on the runway. If you back-taxi down the edge of the runway you may be partially camouflaged, setting the stage for a potential accident.

Crosswind taxi technique is not something that should be touched upon only briefly during flight training. Crosswind principles are important, if not vital, to a student pilot and indeed to every pilot. The concept of crosswind technique is especially critical for students because, being impressionable, they tend to learn better, and use more of what the instructor emphasizes. No one should expect a new pilot to be totally proficient, but he or she should know and be able to verbally describe the principles of crosswind taxis, take-offs, and landings. Understanding this concept is, in actuality, not that difficult when emphasized early in the training and then practiced at every opportunity.

The way I learned crosswind taxi technique is to remember "climb into a headwind and dive away from a tailwind." To do this, first determine which direction the wind is coming from. If it is coming from in front of or parallel to your shoulders, pull the yoke back (climb) and turn the yoke towards (into) the wind. If the wind is coming from behind your shoulders, push the yoke forward (dive) and turn it away from the wind (see Figure 5-3).

Another way of understanding this concept is to envision that you always want to keep the upwind

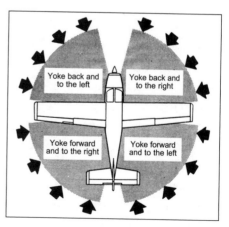

Figure 5-3
Flight control positions for crosswind taxi

aileron in a position where, when the wind strikes it, the upwind wing will be held down. You never want to give the upwind wing any opportunity to rise up. One safety consideration here is to stop and think that if a lot of control usage is needed to taxi properly and safely, it may indeed be too windy to fly.

If you have progressed down a usable taxiway (instead of back-taxiing), you should now be nearing the hold-short line. This line is actually 4 painted lines, 2 solid and 2 dashed. The hold-short line is the delineation between the ground environment and the flight environment. Never cross the hold-short line until you are ready to enter the flight environment. As you approach the hold-short line, turn or cock the aircraft at an angle to the taxiway centerline so that when the engine run-up is performed, the higher RPM will not cause propwash to blow dirt and grit onto an aircraft behind you. Remember, you are not the only one at the airport and a little courtesy will go a long way towards making lasting friendships.

It is now time to review the "Before Takeoff" checklist. This is also called the ground check, the run-up, or some combination of these. The first thing to do is set or hold the brakes. Again here, the C-152 and Tomahawk checklists differ in the order that tasks are performed, so you must follow the checklist for the aircraft that you are flying. I will describe a generalized run-up and highlight certain items to look for.

Double check the cabin doors for closure and security. Set the elevator trim to takeoff position. Re-check the fuel shut-off valve or the fuel selector for the proper tank. Enrich or lean the mixture according to your density altitude conditions. Increase power to 1700 RPM in a C-152 and 1800 RPM in a Tomahawk. At this time, you need to check the magnetos. When you check the magnetos, you are grounding one of them while the engine is running on the other. Turning the key to the left magneto grounds the right one while a check of the left magneto is made, and then vice versa. Do not whip the key too fast or you may turn it to the off position. This action grounds both magnetos, causing the engine to die out. When this happens, fuel continues to flow into the cylinders and, when the key is turned back to the on position, the detonation of excess fuel will explode and manifest as a backfire. This can damage an engine.

Check the magnetos methodically and carefully.

During the magneto check, you must observe the tachometer for a drop in RPM. When an engine runs on only one magneto the fuel burn is less efficient, causing a power loss. There is a maximum drop allowed on any particular magneto, and this may vary from 125 RPM to 175 RPM. One other thing to look for is the difference in the RPM drop between the magnetos. While each magneto could have a 125 RPM drop, there should not be more than 50 RPM difference between the two. This RPM tolerance ensures that the two magnetos are within close timing for firing the engine. Be sure to return the key position to both magnetos when the checks are completed.

The next thing to check during run-up is the carburetor heat. The carburetor heat is actuated by either a push/pull knob in the C-152 or an up/down lever in the Tomahawk. It is irrelevant which actuator you have as the principle is the same for both. Moving the actuator to the carburetor heat _ON_ position moves a wire, enshrouded within a coil to prevent bending, that actuates a flap in the engine compartment. This flap changes the air feed to the carburetor from the normal outside, filtered air to unfiltered, heated air. The heated air starts as outside air which is routed through a shroud that encases the outside of the exhaust pipe or muffler. This outside air does not mix with the exhaust; rather, only the heat from the exhaust is transferred to the routed air. This heated air also can be used for cabin heat.

Activating the carburetor heat allows this unfiltered, heated air to enter the carburetor. Heated air going into the carburetor on demand is a vital tool to have in a plane whose engine has developed carburetor ice, so a check must be done to ensure that the carburetor heat is working properly. To check the carburetor heat, actuate it with the knob or lever in your aircraft and look at the tachometer for an RPM drop. The amount of RPM drop may vary from plane to plane, but you must see an RPM drop. This RPM decrease results from the fact that the heated air going through the carburetor is less dense than normal air, and this "thinner" air causes the engine to run less efficiently. If no drop in RPM is seen, the cable may be frozen or broken.

Do not be complacent on a hot day by thinking that carburetor ice cannot develop. The venturi in the carburetor lowers pressure to pull through the fuel/air mixture, and it also lowers the mixture temperature as much as 50° Fahrenheit overall. Even on an 80° F day, ice can and does form in the carburetor when the temperature inside it falls below 32° F. To takeoff without a properly operating carburetor heat system is to invite disaster.

It is equally important during the carburetor heat check to look for an increase in RPM back to the original setting seen before you activated the carburetor heat. If you do not see an RPM increase, the carburetor heat cable may have just broken, leaving the heat on. An engine with the carburetor heat on will not produce full power on takeoff or anytime during the flight. It is best to taxi back to the ramp if either the initial RPM drop or the subsequent rise in RPM is not seen during the carburetor heat check.

While the run-up RPM is high, check the engine gauges (oil pressure, oil temperature, ammeter and suction) for their proper operating ranges. I will not discuss all of these gauges here because there are so many variations in them and the way in which they are read. Some have yellow, green and red color indications. Some have green only. Some have numerical ranges with which you must become familiar, and some have combinations of these. Whatever the types of gauges on your aircraft, access the POH and be knowledgeable about their ranges and the particular indications to look for before takeoff.

At this point, you should throttle back to low idle and check to see if the throttle cable is adjusted properly. Carefully, slowly and smoothly pull the throttle all the way out or back, depending on your throttle arrangement, and see if the engine continues to run. If it is in proper adjustment, the engine should be at a low idle of approximately 400 to 500 RPM. You can even actuate carburetor heat at idle to see if the engine will continue to run, but do not leave the carburetor heat or throttle in this position for long. Carbon will build up in the cylinders, and you may foul a spark plug. Return the throttle to 1000 RPM immediately after this low idle check.

Unless your checklist has specifically instructed you to do so before this point, it is time to check the flight controls to ensure that they are free and correct. I observe too many people doing this task who either do not know what to look for or who are far too complacent. It is especially important to check the flight controls carefully after any maintenance. If they are rigged backwards by accident, the pilot is in for a dangerous surprise after takeoff. Mechanics are only human and can make mistakes. It must be the pilot who finds the error before takeoff. The FARs state clearly that the pilot, not the mechanic, is ultimately responsible for the airworthiness of the aircraft before flight.

I suggest checking the flight controls in the following manner to ensure that they are free and correct. Place both hands on the yoke with the horns of the yoke and your thumbs pointing up. Turn the yoke to the right. Your thumbs should now point to the aileron that is up. With the yoke and your thumbs still pointing to the right, look to the left side to see that the left aileron is down. It is important to look both ways. Now turn the yoke to the left and ensure that the left aileron is up and the right aileron down. Return the yoke to normal. To check the elevator, pull back and look back. The elevator should be up. Push forward and the elevator should go down. While looking back, push your right foot and look for right rudder; then left foot, left rudder. This rudder check will be more difficult in the Tomahawk because of the direct linkage to the nose wheel. Looking at the rudder counterbalance will help on the Tomahawk, but you must remember that the counterbalance is supposed to move in the opposite direction of the rudder.

Checking the flight instruments should be a methodical process. For a modern panel where the flight instruments are arranged in two rows, such as in the C-152 and Tomahawk, this procedure is easy. Simply check left to right across the top row—airspeed, attitude indicator [AI], altimeter—and right to left across the bottom row—vertical speed indicator [VSI], heading indicator [HI], turn coordinator (see Figure 5-4). Do not just make a cursory gesture or glance at these instruments. It is important to know what to look for in them as described below.

✦ **Airspeed Indicator**—Check to make sure that the needle is on zero. If not, there may be an obstruction in the pitot tube. Such a blockage may have lodged there while taxiing.

✈ **Attitude Indicator** (AI or "artificial horizon")—Look to see that the indication is reasonably level. I use the word "reasonably" because if someone had flown the aircraft shortly before you and then you re-started the engine before the gyro had a chance to spin down, it may actually take longer for it to stabilize again. Also check the setting of the "pip." The pip is the dot in the middle of the instrument. It should be set on the horizon to the preference of the PIC. Be aware, though, that because of unlevel pavement and/or nose wheel tire and strut inflation variables, this pip setting may not be exact. The pip will be at its most accurate when set later at straight and level cruise flight.

✈ **Altimeter**—To properly set the altimeter, you must first understand that the published field elevation is taken from the highest point on the airport. If the published field elevation is 600 feet and this happens to be at the ramp where you initially set the altimeter, whereas the place at the end of the taxiway where you are presently sitting is 20 feet lower, the altimeter should read 580 feet. Do not reset it to 600 feet. If you do, the altimeter is now incorrect by 20 feet.

Figure 5-4
Be methodical when checking your flight instruments

✈ **Vertical Speed Indicator** (VSI)—Ideally, the VSI needle will point to zero on the instrument. However, if it does not, where it does point now becomes the zero reference for climbs and descents. Having it point to zero initially just makes the math easier.

✈ **Heading Indicator** (HI or Directional Gyro, DG)—Now is the time to set the HI. Remember that the HI gyro runs off the vacuum pump, which runs off the engine. The faster the engine RPM, the faster the vacuum pump turns and the faster the HI gyro spins. The faster a gyro spins, the more stable it is. Now that the engine has been run-up to a higher RPM, the HI (and the AI) are more stable and therefore more accurate.

✈ **Turn Coordinator** (TC)—Knowing that the TC is an electric gyro, look inside the TC instrument window face to find a small square window. If the TC is faulty or is not receiving electric power, there will be a red flag inside this small square window. Verify that there is no flag.

At this point, you are very near departure but must perform the final checks. Different aircraft checklists call for verifying various items but, while I emphasize the need for following the checklist for your aircraft, here are a few simple memory aids that might help at this time. The first is the acronym "BLT." I like this memory aid for the Tomahawk. "B" stands for boost pump to be _ON_, "L" for lights _ON_ (including the landing light), and "T" refers to the transponder to be _ON_. This memory aid can also work for the C-152 where the "B" stands for the beacon, the "L" for the landing light, and the "T"again stands for the transponder, which should be on the desired squawk code. For our flight around the pattern at this non-towered field, the transponder code should be set at 1200. Always use the mode "C" function if the transponder is so equipped.

Another acronym memory aid that can be used is "FFLTT." Think of it as "Am I ready for 'flight' (FFLTT)?" These letters stand for **Fuel**, **Flaps**, **Lights**, **Transponder**, and **Trim**. I actually prefer this memory aid because it can be used for nearly any piston airplane, high wing or low, single or light twin engine. What you must be careful of, though, is when you say **Fuel**, you must know to check all the specifics regarding the fuel for that airplane. Is the boost pump supposed to be on or off? Check it. Is the mixture set properly? Is the fuel selector in the right position? Also check the fuel pressure gauge. My rule of thumb is that whenever my hand goes to the fuel pump switch, whether it be for turning it on or off, my eyes go to the fuel pressure gauge for a proper reading.

The same holds true for the **Flaps**. In what position do you want them? Is it a normal takeoff or is it a short field or soft field? A word here about practicing those short and soft field takeoffs and landings on a long, paved, forgiving runway. Think. Think about what you are trying to accomplish. If you are slack about practicing real world procedures, then the real world might seek you out as a victim. Aviation, in general, has a unique way of being very self-cleansing. For soft field practice, really think about and realistically simulate keeping that nose wheel out of the mud. Stay off the brakes also to keep from causing the nose wheel to bog down. For the short field simulation, back-taxi to the very end of the runway so that every last inch is used. I remember practicing for football in high school many years ago. The coaches encouraged us to practice like we would play. Good advice even to this day.

Turn on all the **Lights** that you desire. Think of lights on an airplane as recognition lights. The more the better. Some studies have shown that burning a landing light can reduce the risk of a bird strike. Cheap insurance if you ask me. The **Transponder** check has already been described. The **Trim** should now be re-checked for the proper setting. Some aircraft are forgiving of a bad trim setting, some are not.

For those who are fond of fast food, there is also a memory aid just for you. This mnemonic is MCFLIGHT: **M**agnetos on both, **C**arb heat cold, **F**laps and trim set, **L**ights and transponder on, **I**nstruments and gauges checked, **G**as on proper tank and mixture set, and **H**eads up for **T**raffic.

You can use these memory aids, modify them, or make up your own. Remember, though, never do anything that is contrary to the POH procedures. A good memory aid should help the individual to remember certain things that he or she might be prone to overlook or go through too quickly. The memory aid should complement the aircraft's checklist, not replace it. You might choose not to use a memory aid at all. That choice also is fine, but either way you must use the recommended checklist diligently. Remember that takeoffs are optional, while landings are not. This final point in the flight preparation is nearly the last chance available to make a decision that something is not right. If something is not right, actual or perceived, do not fly.

Before you cross the hold-short line, be sure to visually check for traffic. Do not rely on the radio calls or lack thereof to make assumptions about the traffic situation. Just about the time you assume there is no traffic because you have not heard a radio call, a NORDO Piper J-3 Cub will fill your life with yellow. At the same time, do not forget to make your own radio announcement of your intentions to taxi onto the runway. Now is the time to communicate and communicate well. The standard rule of thumb for a radio transmission is to announce who you are, where you are and what you intend to do.

However, at this non-towered field, I suggest a radio announcement such as: "[name of airport] traffic; [Cessna or Piper] [and give "N" number]; back-taxi, [runway #]; [name of airport]." Notice that the name of the airport is given at the beginning of the announcement and at the end. This is sometimes important on a unicom frequency where there are multiple airports in the area using the same frequency. If all has gone well to this point, you are very near the stage of committing yourself to aviating.

"Stay calm. Don't just start doing things.
Reason things through before you act."

—Art Scholl, former Hollywood Pilot and Airshow Performer

CHAPTER 6

Takeoff & Climb-Out

"Stay up on the edge of your seat."

—Scott Crossfield, former Test Pilot

Now comes the big event that makes this business all worthwhile. You are about to take a flying machine into the air. It is a special moment each and every time. All the checks have been completed, and you have decided that the aircraft and flight conditions are safe for your skill level. Take a moment and sit up straight. Remind yourself that bad things can happen and machines can break. You want to be ready for anything. When you decide to leave the ground in a machine, you carry all the responsibility with you.

To cross the hold-short line, add a moderate amount of power until the aircraft starts to move, then reduce power to around 1000 RPM as you did when taxiing. You have just entered the flight environment. Again, try not to ride the brakes. The goal when moving onto the runway is to position yourself so that all the usable runway can be used for takeoff. Do not cut corners. Taxi to the edge of the threshold and act as though you may need all the runway in case of an aborted takeoff. You never know; you just might. Pilots who learn on long, wide runways can be spoiled very quickly. Remember, one of the three things in aviation that do you absolutely no good is runway behind you (the other two are altitude above you and fuel on the ground).

If the taxiway merges with the runway at the threshold, pull straight out onto the runway. If the taxiway does not merge with the runway directly at the threshold, then you should back-taxi. Do not be shy or intimidated by the back-taxi procedure. Once you have crossed the hold-short line, that runway is yours, and you can do whatever you need to do to make aviation safe. I reiterate; do not waste the opportunity to utilize every inch of runway length. You may need it.

Another thing that should never be taken for granted is the center-line. The centerline is the main measuring tool for ensuring that the take-off is straight and that you do not drift to one side or the other while accelerating. Again, wide runways can spoil a person just learning to fly. A smart instructor will take the student to a short, narrow runway at an early time in the student's training. For now, though, when you are on a long, wide runway, you should imagine that the runway where you sit is short and narrow. Be as precise as you can and align the nose wheel with the centerline. Play the game by the rules now. It will pay off later.

After lining up the airplane directly on the centerline with the nose wheel straight, stop momentarily. Again, sit up in your seat and remind yourself that anything can happen, and you will be ready if it does. Move the control yoke in the direction of any crosswind and advance the throt-tle slowly but deliberately. It should take approximately 3 to 5 seconds for the throttle to reach the full power stop. Do not advance the throttle by quickly jamming it forward. Many engine failures occur during power changes. This is because the engine cannot take the sudden change in temperature or because it cannot handle the sudden mass flow of fuel. Either way, this is a potential pilot-induced engine failure that can be easily avoided. Also keep your right hand on the throttle in case it tries to creep back from full power. Additionally, you want to be able to reduce power quickly if something is not right and you must abort the takeoff.

Once you advance the throttle, the forces of torque and spiraling slipstream immediately exert their effect (see Figures 6-1 and 6-2). In the C-152 and Tomahawk, the result will be seen and felt as a pulling to

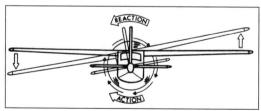

Figure 6-1
Effects of torque (above) and spiraling slipstream

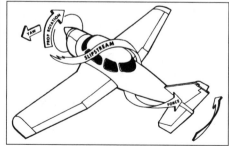

Figure 6-2

the left. Because of the natural tendency to drive a vehicle with your hands, almost every beginner moves the control yoke to the right to counteract the effects of these forces. Early in your training, develop the mindset that an aircraft is steered on the ground by your feet pushing on the rudder pedals, especially during acceleration. Proper use of the right rudder will offset the effects of these forces in these two (and many other) trainer aircraft.

The consequences of failure to use flight controls correctly during takeoff can be illustrated by visualizing a crosswind from the left. Using proper technique, the yoke is turned to the left to put the left aileron up and the right aileron down. This action keeps the left wing down. Now the pilot is free to use the right rudder to counter- act the left turning tenden- cies (see Figure 6-3).

Figure 6-3
Keep the upwind wing down for a crosswind takeoff

If, however, the pilot wrongfully moves the yoke to the right in an attempt to "drive" the air- craft back to the right, countering torque and spiraling slipstream, he or she may get anxious when this has little effect on the aircraft's contin- ued movement to the left. A much larger surprise ensues when, with the yoke to the right and the left wing lifting first, the wind lifts the left wing even more and, in an extreme case, may put the right wing on the ground causing the aircraft to cartwheel, tail over wing. Not a pleasant thought! The scenario is no prettier for a right crosswind, either. Learn to use the rudder pedals early in your training for directional control during take- off. The rudder is the most powerful flight control available to the pilot at slow airplane speeds.

As the airplane accelerates to its takeoff speed, it will get lighter on the ground as the wings generate more lift. It is not wise to hold an air- plane on the ground at too fast a ground speed. Never push forward on the yoke pinning the aircraft to the ground. This makes the airplane effectively "wheelbarrow" down the runway. Wheelbarrowing is very dangerous because directional control is more easily lost. It is also hard on the nose gear. With the airplane already light from lift, the effects of gravity and friction on the runway diminish and the effect of the wind on the airplane's surfaces increases.

I was taught that when an aircraft wants to fly, let it fly. This philosophy has never let me down. During a normal takeoff, glance at the airspeed indicator during acceleration and look for a speed that is 2/3 to 3/4 the best rate of climb speed (Vy). It is near this speed that the nose of the aircraft should be raised slightly even though the whole aircraft is not ready to fly. This action is called rotation. A rule of thumb for rotating the nose for takeoff is to raise and position the top of the cowling at the horizon or tree line and hold it steady there. When the airplane reaches flying speed, it will lift off, usually at or around Vy. There will be no further need to lift the nose or pull the airplane into the air. Basically, what you are doing is putting the aircraft in a position to fly itself off the ground.

When you master this method, you will be pleased at the increased smoothness of your takeoffs. This method will build your rudder use skill and coordination because, with the nose off the ground, the airplane is permitted to pivot on the two main gear tires. P-factor and gyroscopic precession are forces which play a part at the moment the nose is rotated (see Figures 6-4 and 6-5). P-factor will pull the nose of the aircraft harder to the left, so more right rudder will be needed to counteract this effect. And, on two wheels, your need to use the rudder is amplified. Actually, gyroscopic precession on tricy-

Figure 6-5
Gyroscopic precession for tailwheel aircraft when the tail is raised

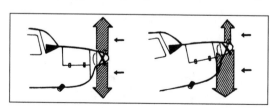

Figure 6-4
P-factor is exerted in a climb (above)

cle gear airplanes will work to your advantage as the nose is rotated up for takeoff. This precession effect helps, albeit minimally, to pull the nose back to the right.

For a crosswind takeoff, this rotation of the nose technique enhances smoothness and confidence. The aircraft will get up on one wheel (upwind wheel), and you will feel more in control. This method requires practice, however, as do all crosswind takeoffs. Do not try too much too fast. An interesting note here is that this position is almost the same one held by the aircraft during landing, except that deceleration will occur during the landing phase of flight.

The moment the airplane's tires leave the runway, you have departed gravity's shoreline and entered an ocean of air. Air is fluid and, as with any fluid, currents are found. In aeronautics, you must compensate for these air currents (wind) to control an aircraft's direction and eventual destination. Relative to the ground, these currents cause the airplane to drift, as a ship on the water would drift with the current. Just as the ship's helmsman must apply steering corrections to compensate for drift, even though he cannot feel it directly, so the pilot must compensate for air currents by observing their direct effect and acting appropriately. A wind which causes sidewise drift is called a crosswind. A plane must never be surrendered to a crosswind and its effects (see Figure 6-6).

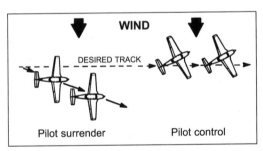

Figure 6-6
Never surrender to the effects of the wind

Some degree of aircraft weathervaning may occur because of crosswinds. Most of us have seen a weathervane turn into the wind. One difference between a weathervane and an airplane is that the weathervane is fixed and rotates directly into the direction of the wind. An airplane moves forward, however, so it will not pivot all the way into the wind. Rather, the nose will rotate into the wind only as much as is needed due to wind conditions and pilot control. Let this weathervane movement happen, because it is normal and even preferred. The goal is to track

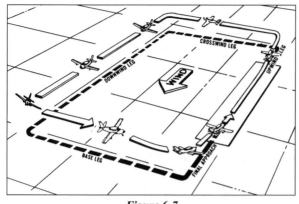

Figure 6-7
The traffic Pattern

straight along the centerline of the runway even after you have left the ground, keeping the wings level. The centerline is your main measuring stick for tracking straight out.

The intentional use of weathervaning for ground tracking purposes is called crabbing (or crab). Imagine a pencil under the airplane. It should scribe an imaginary line directly down to, along, and finally become an extension of the centerline. This initial leg of the climb-out is called the upwind leg (see Figure 6-7).

The importance of tracking the centerline cannot be underscored enough. Here again, imagine a crosswind from the left and envision that, after your takeoff, someone else takes off directly behind you. If you let the crosswind drift you to the right of the centerline, you will turn directly across and into the flight path of the other plane as you turn crosswind (see Figure 6-8). But if you track the centerline straight out, you would turn away from

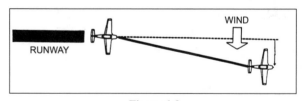

Figure 6-8
An undesirable upwind leg

their flight path when you turn crosswind. Another example emphasizing the importance of tracking a runway's centerline is taking off from a wintry runway where snow banks may be only a few feet from each wingtip. Side drift here just after rotation could be disastrous.

One way to follow the centerline is to use your peripheral vision to follow the edge of the runway where it meets the grass. Keep this line in the same relative position, and wind drift will be minimal. If you begin seeing more runway or more grass, you are drifting sideways and not

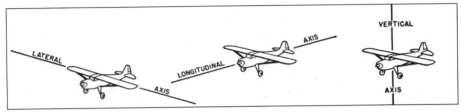

Figure 6-9
The three flight axes of an airplane

tracking straight down the centerline. Practicing the rectangular course, a piece of the traffic pattern puzzle to be practiced away from the pattern, will help reinforce the understanding of this concept.

It should be mentioned here that once you have left the ground, you have left two dimensional travel behind and entered the arena of three dimensions. These three dimensions are roll (turning), pitch (up and down), and yaw (side to side). Roll, pitch, and yaw are utilized in an airplane through the flight axes. Each axis corresponds to an imaginary line through the airplane around which the plane pivots (see Figure 6-9).

The airplane rolls about the longitudinal axis, pitches about the lateral axis, and yaws about the vertical axis. All these axes intersect at the center of gravity. The pilot manipulates the flight controls to command the airplane to behave as he or she wishes. The pilot's hands work the yoke for roll and pitch control, and the pilot's feet work the rudder pedals for yaw control. Three axes for three dimensional travel.

Another very important action immediately after liftoff is attaining the Vy speed for your airplane and holding it (see Figure 6-10). For the Tomahawk, this is 70 knots indicated airspeed (KIAS). For the C-152, it is 67 KIAS. Vy is different for almost every make and model of airplane. This airspeed is import-

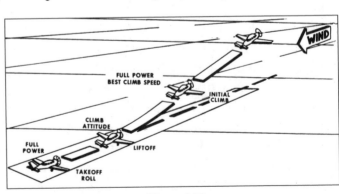

Figure 6-10
A normal takeoff using Vy

ant, though, because it gives you the most altitude per unit of time in the air. This altitude gained translates into more altitude available should some problem occur, such as an engine failure. Do not worry about watching the rate of climb on the VSI. This rate will vary depending on the density altitude and the gross weight of the plane. What Vy will do is give you the most altitude possible for the existing atmospheric and aircraft conditions. Attaining and holding Vy for a normal takeoff is another form of cheap insurance.

Several things can be accomplished on the upwind leg. One is to step on the brakes to stop the rotation of the tires. This action can prevent the tires from consistently stopping in the same position as they wind down, which is caused mainly by an out-of-round condition. When they stop in the same position on each takeoff, they wear in the same spot on each landing. Braking the wheels so the tires stop at random positions after takeoff should prevent a bald spot from developing prematurely. Another desirable action during this early phase of flight is glancing at the engine gauges to verify that all is well with your power plant. Learn to interpret the gauges with a glance. You will get better at this with time and practice.

Assuming now that you have held a good Vy airspeed and tracked the runway centerline, soon it will be time to turn left onto the crosswind leg. Turning left assumes that you are flying from an airfield where the traffic pattern is standard. The place to initiate the crosswind turn is at 500 feet above ground level (AGL). The angle of bank should be about 20° ±5°. More bank than this will make your passengers uncomfortable and increase load factor, while less bank will simply take more time and airspace than necessary. Most pilots are (and all pilots should be) an efficient group of people. Smooth, steady, deliberate, and coordinated flight control movement is the key.

The coordinated turn is something which is fundamental to controlled flight. If a turn is performed without rudder coordination, the airplane will yaw adversely. Notice that the word "adversely" is very similar to "adversary." Allowing adverse yaw to go uncorrected will turn the airplane into an adversary, decrease turning efficiency, and could jeopardize safety. When an airplane is turned to the left, its nose is pulled to the right by the drag produced from the downwardly deflected right aileron. This influence can be seen through the motion of the "ball" in the turn coordinating instrument on the panel. Think of the ball as free floating from side-to-side, able to react to side-to-side motions of the nose. Thus, in a left turn without rudder correction, as the nose is pulled to the right into a condition known as a slip, the ball reacts and offsets to the left (see figure 6-11).

Figure 6-11
A slip during a left turn

The only way to compensate for adverse yaw is proper and timely use of the rudder. It was the invention of the rudder which allowed the Wright brothers to continue successfully with their early glider experiments. Before their rudder invention, turns were awkward affairs, and their glider aircraft were unpredictable during turns, sometimes spiralling to the ground. After a rudder was installed, the Wrights found that the vertical flight control surface area, when properly applied in flight, acted as an aerodynamic lever to pull the nose of the aircraft back to the desired direction of the turn. Left rudder was, and still is, needed in a left turn to prevent the nose from being dragged to the right (see Figure 6-12).

Figure 6-12
Proper use of the rudder during a left turn

For the same reason, right rudder is needed during a turn to the right. The age old memory aid for proper rudder use is to "step on the ball." This simply relates to right rudder if the ball is to the right (right turn) and left rudder if the ball is to the left (left turn). The ball should not move from its center position if the rudder is properly used as a coordinating flight control during turning maneuvers. This does not say that the pilot should look constantly down at the ball during turns. The mechanics of a turn should be understood and practiced until a coordinated turn can be achieved instinctively.

Recall for a moment the story from the "weight and balance" chapter of this book. It was at this stage of the flying game that the pilot lost control and could not recover. Your present situation is different. You should not be in danger of this type of accident if you are within the weight and balance limitations, your airspeed is at Vy, and the bank angle is 20° ±5°. But you will need to compensate slightly for the apparent weight felt by the aircraft because of load factor. This compensation is accomplished by adding slightly to the wing's angle of attack, producing a little more lift, offsetting that apparent weight. This slight increase in angle of attack still should leave plenty of safety margin between the wing's angle of attack in the turn and the angle at which the wing would stall. Turning is an art that must be mastered.

Do not forget to make the radio announcement of your intentions as you turn crosswind. At no time, however, should you try to make a radio call when maintaining control of the airplane is the appropriate priority. Humorously said, do not key the airplane while trying to fly the microphone. The rule of thumb for priority in any airplane at any time is aviate, navigate, and then communicate. Always fly the airplane first and foremost. Never stop flying the airplane for any reason or to accomplish any task.

The airspeed should continue at Vy during this turn to crosswind. What you are accomplishing is a constant airspeed, climbing turn. Roll out of the bank and level the wings the moment you reach a position where the aircraft will track a path perpendicular to the runway. This will not be an attitude where the longitudinal axis will be perpendicular to the runway, unless there is absolutely no wind or a direct 90° crosswind. There usually is wind, though, and you have just taken off into the wind, so making a left turn will put the wind from the right side (see Figure 6-13). Therefore, you can

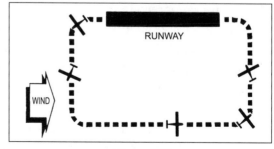

Figure 6-13
The elusive "perfect" rectangular pattern with wind

expect the wind correction (or crab) to be to the right. Remember the imaginary pencil scribing that line across the ground? Now the line should be 90° to the climb-out and/or runway centerline.

Once the wings are level, look to the right to check for traffic. You cannot avoid other aircraft if you do not see them. Always expect traffic in and around the traffic pattern. Look for it. Other pilots entering the traffic pattern generally will use a 45° entry to the downwind. This brings them into the pattern from your right.

Vision is the pilot's most important sense. It is a known fact that most midair collisions occur during daytime VFR conditions and in the traffic pattern. Even more alarming is the fact that most midair collisions are not head-on. They are between two aircraft moving in the same general direction. Clearly, scanning for traffic while in a naturally heavy traffic area, such as a traffic pattern, is vital. Never become so involved with all the cockpit tasks necessary for the approach and touchdown that the scan suffers. The one rule here is see and avoid. You have to see other planes first to avoid them. The pilot should be able to fly the airplane and spend 70% of his or her time scanning for traffic.

Proper scanning requires that the pilot move his or her head and eyes to various sectors of the sky. Every pilot usually develops his or her own scan pattern. This is fine as long as that scan pattern covers all of the sky that can be seen from the cockpit. Sometimes the scan will require that the pilot move his or her head to see around the various blind spots that are unavoidably built into every aircraft. Obviously, the C-152 will have more of a blind area above the cockpit, while the Tomahawk will have more blind area below the cockpit.

To fully appreciate how to scan properly, it is first wise to understand how the eye works. The fovea, the central part of the retina, is where our vision is sharpest. This foveal field of vision is surprisingly small. It is a conical field of view of only about one degree. Now it is obvious that we can see a lot more than just this one degree cone. But do you realize how little detail you can see outside this foveal cone? The truth is that the visual acuity drops off rapidly outside the foveal cone. In fact, outside of a 10° cone concentric to the foveal cone, a person sees only about one-tenth of what is seen within the foveal field. Plug this fact into the flying equation, and we see that if a pilot is capable of seeing another aircraft 5,000 feet away within the foveal field, the peripheral vision likely would never see it at more than 500 feet! Ninety percent is a big loss.

This is why the instructor always says, "Keep your head on a swivel!" This process of scanning for traffic is most important for student pilots who think that they must concentrate on the dash and its instruments to fly. It is extremely important that good scanning habits are developed early in a pilot's career. Many a student pilot has thought that it was crazy for the instructor to use those suction cup covers to block the instruments. Many times it was to get the student's head out of the cockpit to scan for traffic. This is actually another one of those pieces of the puzzle that the student learns outside of the pattern to bring back and fit into the larger mosaic of the traffic pattern.

Rapid and jerky eye movements will not work. Neither will scanning the sky just in front of your aircraft. The pilot must scan the entire horizon along with 10° of airspace above and below it (see Figure 6-14). So the pilot must use systematic eye and head movements to scan small 10° segments of the sky, one at a time. Since the eyes need time to focus within this narrow viewing area, and then possibly refocus from the near vision required inside the cockpit, the FAA recommends that a pilot use a series of short, regularly spaced eye movements allowing at least one second for each segment. With this information, the pilot can begin to develop his or her own scan pattern taking into consideration his or her own visual strengths and weaknesses. Just make sure that your scan is

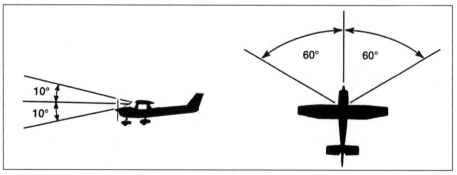

Figure 6-14
Typical traffic scan fields for most aircraft

efficient enough to cover all the sky that can be seen from your cockpit.

If you are a pilot who is required to wear glasses while flying, you will be wise to carry an extra set for backup. Some pilots accomplish this by wearing prescription sunglasses while carrying their clear lens glasses as well. In any event, when you are sure that there are no other aircraft to interfere with your intentions, continue turning to the left into the downwind leg. Make your radio call as you turn to the downwind.

At this time, the same angle of bank and Vy airspeed should be used as with the first climbing turn. One nice result from holding the appropriate Vy airspeed throughout the climb-out and climbing turns is that you should be at or near the traffic pattern altitude of 1,000 feet AGL once you have neared or completed this turn to downwind.

You should wait to be in a position where the aircraft is at least 1/2 mile from the runway before turning downwind. At the same time, you should be no farther than 1 mile from the runway for the downwind leg. This is the appropriate distance for a traffic pattern in a trainer aircraft. Judging this distance is easy when you use the runway itself as a measuring tool. If you just took off from a 5,000 foot runway, you should be no farther from the runway than its linear length. Neither should you be closer than half its length. If the runway is 2,500 feet, then on downwind you should be no farther away than twice its linear length. Adjust your judgment accordingly for any other runway. Personally, I like to teach my students to fly at 1 mile from the runway. This allows more time to think and to judge during the base leg.

Once the nose is lowered to hold pattern altitude (this can be accomplished in a turn so your wings may or may not be level at the time), do not be quick to reduce power. In fact, it is better to lower the nose with full power for a few seconds while you build forward airspeed. The plane probably will want to climb as it builds speed, so a little nose down trim may be needed, depending on how strong a force you like or need to feel in the control yoke. I personally do not fly using a lot of trim in trainer aircraft. I prefer to teach pilots to fly the airplane with what he or she feels in the yoke, using trim as an assistant. If you spend too much time achieving perfect trim, you can get behind in your duties. When you do reduce power on the downwind, reduce it to around 2200 to 2300 RPM. This power setting should work well for the C-152, Tomahawk, or any other trainer airplane.

Now that you understand the principle of wind drift correction and crab, use this knowledge to set up a ground track that is parallel to the centerline of the runway. Think again about the imaginary pencil under the airplane defining a straight line across the ground. It is important to set up any crab for crosswind early in the downwind phase, because then you have the runway as a reference line tool. You will not have that reference when you pass the other end of the runway, yet you should be able to hold the parallel track. At that, abeam the threshold, you will be busy with other things preparing for the landing. That is no time to waste time trying to hold a parallel ground track. When you set it up early, it is easier to hold this ground track subconsciously as you manipulate the plane into position and its controls into their positions for the landing phase.

"Does it pass the common sense test?"

—U.S. Air Force Thunderbirds

CHAPTER 7

Downwind, Descent and Landing

"Keep your brain a couple steps ahead of your airplane."

—Neil Armstrong, former Test Pilot and Astronaut

The time to start planning the landing is before you enter the downwind. In fact, most good landings are set up before the downwind begins. This simply means that a pilot should have a good mental picture very early in the flight of what tasks lie ahead. Early in the downwind, though, use your "Before Landing" checklist to ensure that all preparations are completed for the landing. The C-152 POH calls for three things:

1. Seats, Belts, Harnesses ADJUST and LOCK

2. Mixture RICH

3. Carburetor Heat ON (apply full heat before closing throttle)

I would add to this list that you should check the fuel selector valve for the *ON* position and turn on the landing light for anti-collision reasons. Just as in some states where motorcyclists are required to burn headlights because it makes them easier to see even during the daylight hours, an aircraft is also more likely to be seen with its landing light burning. So it makes sense to turn on the landing light. Reducing the risk of a bird strike is one thing; being seen by another aircraft has a much larger payoff. Again, this is a cheap form of insurance. One note here: if you have a separate taxi light, you might use it instead of the landing light. Saving as much of the landing light as possible for night flight is smart because landing lights do have a notoriously short life span.

The Tomahawk pre-landing checklist calls for extra actions appropriate for that aircraft. In addition to two of the above three items called for in the C-152 (the Tomahawk checklist does not mention applying carburetor heat), Piper specifies that the fuel selector be put on the proper tank, which is usually the most full tank, and that the electric fuel pump be turned on. Even if you know the fuel pump is on, double check it as well as the fuel pressure gauge. Although the Tomahawk checklist makes no mention of using carburetor heat, I recommend that you use it in that airplane. Stay at the traffic pattern altitude of 1,000 feet AGL or as recommended in the "Airport Facilities Directory" and keep your ground track parallel to the runway centerline throughout the entire downwind (unless, of course, deviation for traffic becomes necessary). One way to help define the parallel ground track is to pick a landmark(s) at some point(s) along the desired flight path to help visualize a straight line. Then any drift can be recognized and corrected for against this visualized line.

By now the plane has moved forward to somewhere between midfield and abeam your touchdown point. At this time, things get busy for the student pilot. Because the plane is constantly moving and cannot stop to allow the pilot to catch up, the pilot is forced to keep ahead of the airplane. When the pilot is not able to keep up, he or she is said to be "getting behind the airplane." In other words, the plane is moving faster than the pilot's ability to perform the tasks necessary to prepare for landing. Certain tasks must be accomplished at certain times; therefore the pilot must constantly think ahead to whatever tasks come next. The two most important items to consider are the next two. The quote by Neil Armstrong that introduces this chapter gets this message across clearly.

The following procedural question is the one asked of me most often. Students frequently inquire, "Where are the procedures written down that list the sequence of things to do to prepare to land?" I have to tell them that there are no written procedures, because every landing is different. You cannot say that something is done at this point or that place every time, because each set-up to land is unique. I recognize that students need a starting place, however, and there must be some standard upon which to base all other variables. This section attempts to compile such a list, but keep in mind that my explanation of procedures is not definitive and different instructors will many times have different techniques and suggestions. An abbreviated landing procedures list is included at the end of this chapter.

When you reach the key point, where you are abeam your touchdown spot, turn the carburetor heat on and throttle back to around 1,700

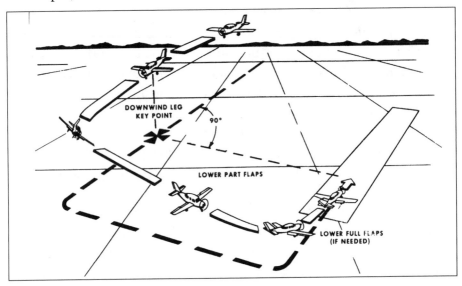

Figure 7-1
Look for the point where the airplane is abeam the touchdown spot

RPM (see Figure 7-1). The plane will try to descend at the moment of power reduction, but do not let it. What you are accomplishing here is an airspeed transition, while holding constant altitude. As the airplane slows, the yoke pressure will build and some minor nose up trim might be desired. I do not recommend trying to trim the airplane to perfection at this time. It takes too much time away from flying the airplane. Simply move the trim wheel one brief roll for nose up and get on with your work.

Now that you are holding a constant altitude with a power reduction, the airspeed will bleed off naturally. In fact, at a constant altitude, the airspeed will bleed off approximately 7 to 8 knots for each 100 RPM of power reduction for a typical trainer airplane. If you do not reduce enough power or if you let the nose drop and lose altitude, the airspeed will not bleed off, and you already will be behind the airplane. The goal in a "perfect" pattern is to reach an airspeed where the first notch of flaps can be applied while still approximately at traffic pattern altitude. For all fixed gear trainer aircraft that I know, the airspeed you want to see is the top of the white arc (the white arc being the flap operating range). While it is important to know what the exact maximum numerical speed is for extending flaps (Vfe), at this time you need not focus on that particular numerical speed. Simply use the color coding on the airspeed indicator. When you first reach the white arc, it is usually time to add the first notch of flaps.

At this point in the approach, the addition of the first notch of flaps is generally accepted as standard, but it must be said that adding flaps is never automatic at any time or position in the traffic pattern. Anyone who has learned to fly by a rote method such that he or she can perform tasks only at certain times or positions should find another instructor to undo this damaging education. The landing sequence is too dynamic to learn as a "monkey see, monkey do" activity. Remember, you are a thinking, acting, and reacting pilot. It is more important to build good judgment so that you can manipulate controls and control surfaces when actually needed, thereby adjusting for varying traffic situations, wind, density altitude, aircraft, and runway conditions.

Adding flaps causes the aerodynamics of an airplane to do two things: add lift and add drag. The flaps will provide more lift to help you stabilize at the slower airspeed and they will lower the stall speed. They also will add drag to help slow your airspeed while in level flight. In other words, they act as a speed brake. Now you can start thinking about descending, but only when you have reached the desired airspeed. Once you start your descent, the flaps help by allowing a steeper descent without increasing your airspeed. For the Tomahawk, I recommend an initial descent speed of 77 KIAS (if it has the inboard and outboard flow strips installed). For the C-152, I recommend 70 KIAS. It will be at one of these airspeeds that you will initiate a constant airspeed descent.

If there is one thing my students will tell you, it is that I preach airspeed, airspeed, and airspeed. I like to say that airspeed is life and control of airspeed is control of the quality of life. Too much is not good and too little is not good. As a pilot, you are best served when you are at the right indicated airspeed. Since performance is speed critical, airspeed management is one of the principal keys to good approaches and landings. In fact, airspeed control is the single most important factor in achieving landing precision. Every single judgment decision you make in the entire approach regarding altitude (too high or too low), use of power (addition or subtraction of it), and addition of flaps (to add or not) should be based on first being at the right airspeed. In every airplane that you ever fly, get to the right airspeed and stay there. Poor airspeed control is detrimental to all else you do, especially when preparing to land.

One point that I will emphasize here is never, ever freeze on the controls with indecision. If you are a pre-solo student and have a tendency to be indecisive and hesitate at the controls, clearly you are not ready for solo flight. If you are a solo or a private pilot and you have a moment of indecision, reluctance, hesitation, or loss of confidence, make at least a go-around decision. Then regain your composure and confidence and try it again. You already have made a good decision by going around for another try.

As you move further past the runway threshold at a constant indicated airspeed, aircraft configuration, and rate of descent, it soon will be time to turn onto the base leg. Once again, where to do this varies depending on the wind and flight conditions. With con-

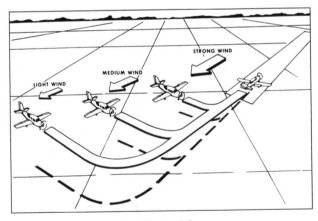

Figure 7-2
The strength of the wind will determine where a pilot turns on the base leg

stant power, a stronger wind will require turning sooner; a lighter wind, later (see Figure 7-2). If you are number one in the pattern for a landing, the textbook no-wind position at which to turn base is when you are at a 45° angle from the runway as defined by a line from the threshold to the base key point and the extended runway centerline (see Figure 7-3). This position may not always be the ideal, however, so remember to

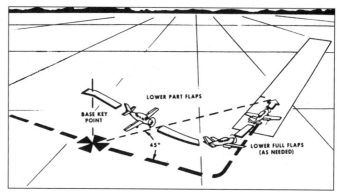

Figure 7-3
The conditional "ideal" point at which to turn base from downwind

plan ahead. If you wish to track the base leg at a certain distance from the threshold, (for example, at a line perpendicular from the extended centerline to the 45° point), you cannot begin your turn at that 45° point, because you will track a line further than desired from the threshold. You must plan ahead and begin the turn prior to this point, especially if you have a tailwind, which will increase your groundspeed and thus require a higher rate of turn.

When you are on downwind and if there is a plane ahead of you in the landing pattern, allow that plane to pass behind your wing before turning onto the base leg. This will permit a safe following and separation distance (see Figure 7-4). Let us suppose, though, that proper planning and judgment were exercised, there is no con-

Figure 7-4
Allow preceding traffic to pass before turning base

flicting traffic, and you have begun your turn to base with little or no wind. As before during climb out, a bank angle of 20° ±5° is suggested for this turn. Be sure to hold your constant indicated airspeed through this turn. The plane will try to nose down a little. Do not allow this. Nosing down will increase your airspeed. You should anticipate this reaction and add a little back pressure to the yoke during the turn to compensate for the loss of vertical lift. The goal here is a constant airspeed descending turn executed very smoothly. Do not "bank and yank" or you will startle your passengers, increase margin for error, decrease safety, and diminish your efforts to fly well.

Here, once again, you must consider the wind. If there is no wind, then it is easy to continue the approach by rolling the wings level just as the longitudinal axis becomes perpendicular with the extended runway centerline. If, however, wind is present and blowing down the runway on which you plan to land, you must consider compensating by rolling the wings level with the nose pointing slightly in toward the runway. This is the same kind of correction that you made on the crosswind leg, except here you will make a turn greater than 90°. This compensation prevents drifting further out from the threshold. If you do not compensate for wind you will drift farther out and take longer to land, necessitating a demanding judgment adjustment. Consider also that if you do track the preferred rectangular course, you will not hold up any traffic behind you.

At this point on base leg the wings are level, the airspeed is still relatively constant and the airplane is moving toward the extended centerline. If you are in a C-152, this is the usual place to add the second notch of flaps. If you do decide to add the second notch of flaps at this time, throttle back about 200 to 300 RPM before you do so. Your approximate tachometer reading after throttling back this second time should be around 1400 RPM. If power is not reduced before adding flaps, the result probably will be an uncomfortable rising of the aircraft. I have heard one instructor compare this situation to putting on the brakes while one foot is still on the accelerator. Throttling back before you add flaps will prevent that upward surge, which is awkward to a beginner and uncomfortable to passengers you may carry after you obtain your license. Remember, though, that adding flaps should never be automatic at any time or place in the pattern, and that reducing power slightly before you do add flaps will make the transition smoother.

In the Tomahawk, the same power reduction should probably take place at this point during the base leg, but no flaps should be added. In either aircraft, a 5 KIAS reduction should be achieved. Now, for the Tomahawk, you should be approximately at 72 KIAS, and in the C-152 at 65 KIAS. Trim again, if necessary, but do not trim too much. The aircraft is still moving, and you have little time before turning from base to final. Spend more of your time on airspeed control and issues of good judgment, such as when to begin the turn onto final approach. Do not forget to look for traffic before turning final. Never assume that you are the only one in the pattern. Remember, there are plenty of folks out there who fly with no radios, and they sometimes appear out of nowhere. VFR conditions mean you should see and avoid. You cannot avoid other aircraft if you do not see them. You cannot see them if you do not look for them.

Now let us assume that you are lined up on the runway centerline after having employed the proper planning, judgment, angle of bank, and smoothness in preparing and executing the turn from the base leg to the final approach leg. The two main things that I stress as requiring concentration after turning final are airspeed (there is that theme again) and keeping ground track on the runway centerline. If airspeed control has not been a problem to this point, the tasks ahead are much easier. Although airspeed and centerline are not the only two items on which to concentrate, they do establish a solid foundation upon which to build the rest of the approach. Only a glance with minimal head movement is required to achieve the division of attention needed for airspeed and centerline attention. This process also acts as a good catalyst for developing the vital art of dividing your attention between inside and outside the cockpit.

If you are not on the centerline, get there as soon as is practical. Do not perform gross maneuvers, mild aerobatics, or small miracles to get there, but be aware that the sooner you can establish the centerline, the sooner you can establish a stabilized approach. The centerline is your aiming device. Even though the nose might not point straight down the middle of the runway, it is vital that your ground track is straight down the middle of the centerline. Whether or not you stay on the centerline will depend on the wind and how you compensate for it. When there is a direct headwind down the runway, the tasks and landing ahead will be fairly straightforward. When the wild card of a crosswind is thrown in during final approach, the planning, preparation, judgment, and decision-making all intermingle and you can become very busy, mentally and physically, right down to the ground.

There are three methods to compensate for a final approach crosswind: (1) the crab method, in which the longitudinal axis of the airplane is not parallel to the runway and this attitude is held right down to the round-out; (2) the side-slip method, where the rudder is used to align the longitudinal axis with the centerline and the yoke is used to lower the upwind wing, thus preventing side drift; and (3) a variation which incorporates a combination of the previous two methods. Every instructor teaches slightly different techniques. I like to teach and use the crab method, and I will tell you why.

I learned about crosswind approaches the hard way while building cross country time. On one flight, I had my wife on board. We were traveling to Hilton Head, SC, to see her family, and all was going well. Landing to the north, I turned final on the runway centerline, which is parallel with the beach. This revealed a strong crosswind blowing from my right, off the ocean. I used the side-slip method and was coming in for what I expected would be a good approach and landing. As I got to 20 or so feet above the runway, I suddenly lost all the crosswind for which I had set up so beautifully, because a tree line now separated the ocean wind from the plane. In other words, the wind at ground level was different than that which I had prepared for. It was a swirling wind. I assure you that this was no place to be scrambling for the correct control inputs. We landed noisily (use your imagination), and it was very uncomfortable, but we were on the ground without any physical damage.

That day, I learned that it does not matter what the wind is doing at 20 or more feet above the runway. The wind that really counts is from 20 feet down to ground level. I could have tracked the centerline just as well using the crab method and waited to use that extra brain power for the proper control inputs when I really needed them. No matter which method you choose to use on final approach, remember that the two things that become paramount to the pilot on final approach are airspeed control and tracking the centerline. Use the same mental tool of visualizing the pencil under the airplane that transcribes a line directly on and towards the centerline and then divide your attention between that line and the airspeed indicator. Airspeed and centerline should become the main, although not the only, division-of-attention items during the approach after turning final. All other judgment will be keyed off these two items and the rest of the approach will require that they be within close tolerance. Once you are within 20 feet of the ground and near the landing, however, the most important focus becomes keeping the longitudinal axis of the airplane parallel to the ground track, no matter which technique you use.

The stabilized approach that I mentioned before is another important element of a good landing. "Stabilized" is defined as a constant airspeed, consistent (straight) ground track, and a constant rate of descent. The rate of descent is predicated on how much power is left in or reduced at the desired constant airspeed. If the airspeed is not constant, all bets are off and another wild card is thrown in. With too many wild cards everything starts getting sloppy, because you become loaded down with unnecessary decisions. Therefore let us presume that the indicated airspeed is, and will remain, correct and constant. This being the case, if power is left constant, a constant rate of descent will occur. If power is added, a lower rate of descent will result, and if power is reduced, a faster rate of descent will occur.

The strategy is to reduce the power enough to descend to a predetermined spot that you have selected, at a constant airspeed, while retaining enough power to pull you to the runway. One way to tell if you are on a proper glidepath to your predetermined landing spot is to keep that spot in the same position in the windshield at all times. Look for a bug spatter or scratch to help fix a position on the windshield. Most pilots use the middle of the runway number(s) as the most popular spot at which to aim. If the spot creeps down the windshield, as if going under the nose, reduce power because you are too high. If the touchdown spot

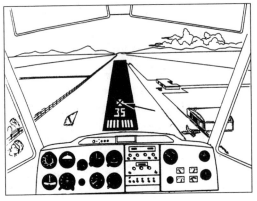

Figure 7-5
Using power changes, keep the desired touchdown spot in the same position on the windshield

moves up the windshield, add power because you are getting too low (see Figure 7-5). Once again, the success of this method is dependent upon constant airspeed. This technique, however, also is dependent upon minimal vertical movement of the pilot's head and/or low turbulence conditions. Under most circumstances, with good planning and judgment, a large or drastic power change should not be required during a normal approach. Normally the descent rate should not exceed 400 to 500 feet per minute, as well.

The decision to add the last notch of flaps on final approach is discretionary, as are many other steps of the landing process. I recommend adding the last notch of flaps only when a landing on the runway is assured and never in doubt. If full flaps are added too soon, you risk getting behind the power curve, where too much drag is added too soon. With this scenario, added power is required to overcome the additional drag. You end up with high drag, high power, and low airspeed. This is not a good situation. If you do find yourself in this predicament, make a go-around decision and get out before it develops into a dangerous situation. The best advice I can give to avoid getting behind the power curve is to wait and add the last notch of flaps only when the runway is assured and, if possible, when your wings are near level.

One note here: the power setting that I last recommended on final approach (1200 to 1400 RPM) is an approximate "zero thrust" power setting. Zero thrust is that power setting at which the propeller is providing neither thrust nor drag. Once the power is reduced below this point, a fixed pitch prop will act as another speed brake and, along with full flaps, will slow the aircraft significantly. I teach my students first to make the decision that the runway is assured and then to reduce all power to idle before applying the last notch of flaps. Nevertheless, if your airspeed management and centerline track have been good up to this point, almost any technique you use should reduce the number of tasks that must be considered on short final and round out.

To avoid a too low airspeed situation, after reducing all power and/or adding full flaps, you must anticipate lowering the nose slightly. Remember that more flaps make for a steeper descent at a constant airspeed (see Figure 7-6). I will grant that flaps also act as a speed brake and help to reduce the extra 5 knots you need to lose to attain short final speed. However, unless you lower the nose, the flaps ultimately will cause an undesirable diminution of airspeed.

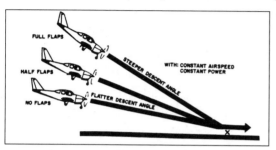

Figure 7-6
The more flaps you use, the steeper the descent at a constant airspeed

This is part of the "feel" or "touch" that every pilot must develop with the "pitch and power" treatment of an aircraft. After adding full flaps, you aim for an airspeed on short final of around 67 KIAS in the Tomahawk and 60 KIAS in the C-152. You should maintain this airspeed right down to the round-out. In fact, you must bring no less than the minimum correct airspeed into the round-out because you must have some reserve airspeed to work with for the landing. Another wild card is dealt to the pilot if there is too much or too little airspeed at this stage of the landing process.

I am a big proponent of using full flaps on every landing, even crosswind landings. Full flaps make for the slowest stall speed and allow the slowest possible touchdown speed. The aircraft will be easier and safer to control as it rolls down the runway. Less tire and brake wear will result, as well. A good friend of mine who works for the FAA advises, "If you are going to crash, crash as slowly as possible." This is good advice, especially if you consider that after all is said and done, landings are in actuality only controlled crashes.

Now comes the anticipated transition that is frequently troublesome to novice aviators: the round-out. Almost all new aviators and even some more seasoned ones suffer from what I call ground rush. Ground rush is that point in the approach where the ground appears to be rushing up to meet you at an alarming rate. It is disconcerting at the least and dangerous at the worst. This sensation develops in the last few split seconds of the approach and can be very dangerous if the pilot becomes fixated on it.

The way to avoid ground rush is to learn to look out the front cockpit window and down the runway rather than fixate on the ground right in front of the aircraft. Focusing on the ground immediately ahead blurs vision (see Figure 7-7), and fixation can result in flying the nose gear (or worse) into the ground after hesitating too long before the round-out. Fortunately, this actually happens only in rare instances.

Figure 7-7
Focusing too closely in front of the airplane blurs vision

Looking at the far end of the runway is not good either. It is best to direct your sight to a shallow downward angle of 10° to 15° toward the runway (see Figure 7-8). It is important to reassert at this point: never freeze on the controls and allow things to just happen without your input. If you feel this taking place, then, at a minimum, add power and exercise a go-around. Keep thinking, and keep your body moving.

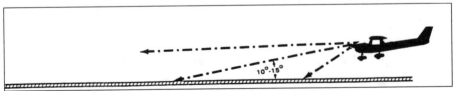

Figure 7-8
During round-out, look out the front window at an angle of 10 to 15 degrees

As with everything else in the landing phase of flight, thinking ahead helps with the round-out. Before you get to the point of looking out and down the runway, ask yourself how you want the airplane to end up. My answer is straight and level, with the wings roughly ten to twelve feet above the runway. I say "straight and level" because, if there is no wind, you will end up basically straight and level with full flaps and little or no power. It is here that the wings will feel the cushion called ground effect. Ground effect occurs roughly when the airplane is within 1/2 the plane's wingspan above the ground. Obviously you cannot land in the short final approach attitude with the nose pointed down. At a minimum you want the aircraft in a straight and level attitude. Of course, if there is a crosswind, you must use the rudder to align the longitudinal axis with the centerline and use aileron input to prevent side drift (see

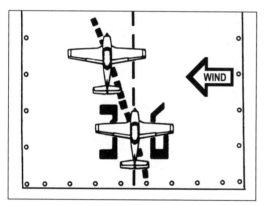

Figure 7-9
Side drift during round-out is dangerous

Figure 7-9). But at some point, you have to decide when and where to raise the nose from the approach attitude to a position where the longitudinal axis is near parallel to the runway surface.

While I teach my students that they should end up 10 to 12 feet above the runway in "straight and level" flight within that precious window of airspeed (not too fast, not too slow), I should qualify this. In the C-152, if your wings are 10 to 12 feet above the runway, your eyes are actually roughly eight to ten feet above it. The landing gear is then approximately three to five feet above the runway. The decision of where to start the round-out is not exact and this judgment is one of the most difficult to teach and learn. Once again, this scenario is indicative of the "touch" you must develop. Only through practice can the pilot's judgment be refined. Visualize where you want to end up after the round-out, then put the aircraft where you want it. However, you must know where the ground is and its relationship to you at all times.

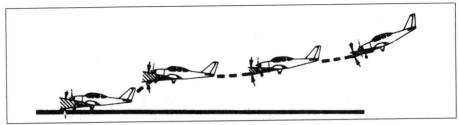

Figure 7-10
Too high a round-out will result in very hard landings

Be careful not to round out too high. Do this, and you can run out of airspeed while still out of ground effect (see Figure 7-10). Ground effect is a cushion of air caused by the ground's interference with airflow around an airplane in flight. Most ground effect results from airflow about the wings, which then reacts with the ground. Even in the early days of flight, pilots found that as they neared the ground for a landing, the ground appeared to push back. You may want to think of it as an invisible pillow of compressed air that buoys an airplane. Whichever explanation helps you, know that ground effect is airspeed dependent and that managing airspeed is managing energy. You must bring some airspeed into the landing and into ground effect. If you do not, you will effectively bleed away your airspeed too soon and have no energy to work with as you get closer to the ground. You will sink right through what should be ground effect. Stated another way, airspeed is the paint and the landing is the canvas. Without enough paint, you cannot complete the picture.

I have repeatedly mentioned airspeed management, and now is when that management pays off. Carry too much airspeed (energy) into the landing, and you float far down the runway, using too much of it, allowing the wind more opportunity to jostle you around. This extra floating may give you more time to become nervous with the landing and thereby more time to make mistakes. On the other hand, carry too little airspeed (energy) or waste it away, and you risk sinking through ground effect too quickly and landing too hard. I must emphasize again that sometimes the best decision is a go-around when you see that things are not proceeding smoothly for a landing.

If you feel that you must add airspeed to any landing, do so with

BEATRICE PUBLIC LIBRARY
BEATRICE, NEBR.   68310

this rule of thumb: add 1/2 the gust factor. If the wind is reported as 10 knots, gusting to 15, the gust factor is 5 knots. Add 2.5 knots to your approach speed. This may not sound like much but consider that a 10% increase in landing speed likely will result in 20% more than normal runway used. If a 20% increase is added to your normal approach speed, you will leave as much as 44% more of the runway behind as compared to what normally would be required with the correct approach speed.

Once you have maneuvered the plane into a "straight and level" attitude, with good airspeed and with the nose pointed in the direction of travel (parallel to the centerline), all that remains is to wait for the plane to sink into the cushion of ground effect. As it does, you must add some up-elevator to raise the nose. This action will increase the angle of attack of the wing, giving it more lift. Holding the nose there will decrease the descent momentarily until more airspeed bleeds off. As the airplane loses airspeed and sinks again, increase the angle of attack to slow the descent again. Before you know it, you will be on the ground with the main wheels first. On every landing with a tricycle landing gear airplane, the main wheels should touch first (see Figure 7-11). If you had already put the plane in a basically "straight and level" attitude during the round-out, any movement you add will cause you to land with main wheels first. The trick is to avoid adding too much elevator too soon, causing a ballooning effect. This is another difficult concept to teach and learn and is part of the "touch" that every pilot must acquire. With practice, though, this process does become easier. Every pilot has to feel the plane in his or her own way to make the proper judgmental decisions.

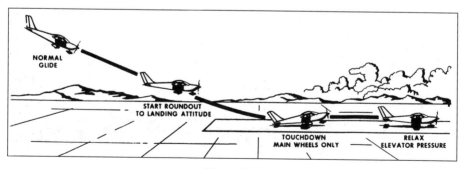

Figure 7-11
Landing on main gear first protects the nose gear, which is more prone to damage

For crosswind landings, use the rudder first to align the nose and fuselage with the centerline. Then, nearly simultaneously, you should lower the upwind wing with aileron input to prevent drifting sideways. If you use aileron without the rudder, you are driving the plane as you would drive a car, trying to steer the nose with your hands. You must remember that the plane is not a car. It is an airplane. You cannot drive it like a car. You must fly it, and flying an airplane requires that you use your feet to move the rudder. The rudder continues to be the most under-utilized and misused control surface on any airplane. Learn to use it properly and effectively. This probably is a good time to mention that the single most frequent cause of aircraft accidents on landing is adverse winds. It is wise to learn your crosswind landings and learn them well.

One technique used by many instructors to teach crosswind landings is to actually touch down only after the first few approaches. The student first is asked to fly down the entire runway length, on centerline, at an altitude of 50 to 100 feet AGL and is told to allow the plane to crab while concentrating on the centerline ground track. On the next low approach, the student is urged to align the longitudinal axis with the centerline using rudder only. This will result in a side drift to the downwind, and the student should see this. On the third pass, the student is allowed to align the fuselage with the centerline and then include aileron input to control side drift so that the centerline is now tracked with the longitudinal axis above and parallel to it. The aforementioned passes may have to be repeated several times in variations. Finally, on the last pass, the student is allowed to reduce power, exercise the skills in whole that he or she has just practiced, and actually land. With this technique, students should find that they are more comfortable with the control inputs for a crosswind landing even though they now have to steadily increase rudder and aileron input as the airplane slows to compensate for decreasing control surface effectiveness. Remember, the most powerful flight control at slow airspeed is the rudder.

After the main wheels finally touch, do not be too quick to put the nose on the ground. Too many times I have seen a person make a good main wheels first landing and then suddenly relax all back pressure, sometimes even pushing forward on the yoke to wheelbarrow. This can be dangerous if you try to drive the nose on at too fast an airspeed or in a crosswind. Instead, try holding the nose up and off the runway as long as possible, controlling the airplane with the rudder. Holding the nose up as long as possible utilizes what is known as aerodynamic braking. Aerodynamic braking is drag from the underside of the wings, fuselage and flaps. Utilizing this will help to dissipate ground speed and save wear and tear on the brakes, when runway length and time allow. Do not pump the brakes after landing. This will produce hot spots on the brakes. Instead, hold constant pressure appropriate to the type of landing that you are trying to accomplish.

Only after the airspeed drops to a safe level should the nose wheel be allowed to come down. Essentially the nose wheel should touch the runway only after all residual lift energy is dissipated. If all has gone well with the landing, the nose wheel will contact the white centerline stripe. Most of us have seen NASA's space shuttle land. If you have not, watch it at the next opportunity with a discerning eye. Shuttle landings are near perfect examples of proper landing technique, where the main wheels touch down followed by a gradual and gentle lowering of the nose wheel.

The best time to practice landings is early in the morning. The wind is usually calm and traffic is usually light. The next best time to practice landings is late afternoon. The winds are usually calm then, too, but traffic may be heavier. For practicing crosswind landings, first select a runway that has a moderate yet steady crosswind. Do not select a day where the crosswind is shifting and/or gusty. It will be too difficult to teach or learn. The crosswind should be strong enough to clearly illustrate the lesson, yet should allow a good safety margin within the limitations of the plane and pilot.

As you probably have surmised, the descent, approach to land, and landing all present you with a constant barrage of decision-making. Utilizing proper throttle control to lower you to the desired spot, adding flaps only when needed, bringing the proper airspeed into the round-out, adding rudder, aileron and elevator inputs—if you do all this correctly, you can land just about anywhere. But do not be too quick to overlook the physiological process that is really going on here. The eyes must first see. The body must first feel. They both provide input to the brain, which makes a decision. The brain then communicates to the musculoskeletal system to move in a certain way. The body appendages supply input to the engine and flight controls, and the aircraft finally reacts to the initial stimulus seen or felt. This chain reaction must happen in a split second. But as you learn to view and feel the aircraft as a part of yourself instead of as a separate entity, you begin to transition to a creature of the air as opposed to a creature of the ground.

If you are a pre-solo student pilot, put yourself in the instructor's position for a moment. Your instructor has to help you understand, and eventually perform, something that you have never done by yourself. Think of that first basic simple landing as a sweet, juicy, exotic fruit that you have never tasted. Now put the shoe on the other foot. If you were in a position where you had enjoyed perfectly ripe bananas for many years and your instructor had never eaten one of those soft yellow things in his or her entire life, how would you describe the wonderful, unique taste of a banana? Think about it. You would have to describe something very abstract to him or her, yet very familiar to you. This is the instructor's quandary at times. Now perhaps you can begin to sympathize with the difficulty of the instructor's job. And with the variety of people who come to your instructor to learn to fly, he or she has to describe landings like a peach to this one, like a pear to that one, and perhaps like a banana to you. Yet not one of you will know how the fruit truly tastes until you finally bite into it.

Still, there is so much more to landings than the basics that I have covered here. You must learn to recognize the visual illusions associated with the length and width of different runways, weather conditions, night operations, speed and glideslope illusions, and many other factors. Additionally, there are very different operating techniques between standard sea level and high density altitude conditions. For these reasons, you should be current and proficient with the procedures necessary for your intended operations and be very cautious when making an approach to an unfamiliar airport. However, the process of evaluating a certain set of conditions for a landing can become extremely efficient, so efficient, in fact, that the pilot may not even be aware of the process that is taking place in his or her brain. But this high level of competency can be achieved only through practice, practice, and more practice.

I recall practicing for my commercial license. The old runway 27 at Stallings Field in Kinston, NC, had a usable length of 2,280 feet with tall pine trees at the far end. I was in a Cessna 150. It was January, so the performance was good. I ended up practicing short field takeoffs and landings for over two hours, with 25 of them logged. The tower got so accustomed to me that they cleared me to land shortly after takeoff. This concentrated practice helped refine my skills to good commercial pilot proficiency, with short field procedures and technique, before transitioning to the more complex Cessna 172RG (retractable gear). The real payoff came a little more than a month later when I passed the commercial checkride. Short field operation, among the other required tasks, had become a sharpened tool that I applied with confidence. I cannot stress it enough—practice, practice, and more practice.

So, as it is inevitable that every student I train eventually asks for an abbreviated written description of the landing procedure from the downwind to touchdown, here it is. I remember asking for one myself from my "Coach" when I was a pre-solo student. This list is abbreviated, but it is as complete as I can present it here. Use it as a guide, but remember that all landings are inherently different in some way. You have to read between the lines and use the gray matter between your ears. Keep in mind that this description may not apply to all aircraft or all conditions of flight. This description is neither intended to replace nor take precedence over any manufacturer's instructions or recommendations. It is entirely possible that you can take this list and modify it for your airplane and purposes.

1. Use the pre-landing checklist.
2. Abeam the touchdown spot, throttle back to 1700 RPM.
3. Make an airspeed transition to the white arc and add the first notch of flaps.
4. Allow airspeed to decay to 10 KIAS above final approach speed and start descent at this airspeed.
5. Keep ground track parallel to centerline and when 45° to corner of runway, turn base.

6. Keep airspeed constant during 20° banked coordinated turn.

7. Roll wings level, throttle back to 1400 RPM, and add second notch of flaps (if aircraft has 3 notches).

8. Allow airspeed to decay to 5 KIAS above final approach speed and track perpendicular to the centerline.

9. Check final approach for conflicting traffic and, if none, turn onto final approach.

10. Establish final approach by centerline ground track, constant airspeed, and constant rate of descent (the stabilized approach).

11. Keep airspeed constant at 5 KIAS above desired final airspeed.

12. Adjust power as necessary to keep chosen point of touchdown in fixed position of windshield.

13. When the runway is assured, add full flaps and reduce all power.

14. Adjust elevator as necessary to keep proper airspeed and prepare to transition to level flight.

15. Round out to level flight with your eyes 8 to 10 feet above runway.

16. Use rudder and aileron inputs as needed to correct for any cross-wind.

17. Be patient to feel the airplane sink into ground effect and adjust elevator-up accordingly.

18. With each sink toward the ground, adjust elevator up accordingly until main wheels touch.

19. When main wheels touch, do not let nose down immediately.

20. Use rudder to steer until nose wheel touches runway from lack of airspeed and then brake accordingly.

After the landing is accomplished, however, your work is not complete, and this is no time to get complacent. You must taxi back to the parking area and secure the aircraft.

"Maintain thine airspeed, lest the ground
riseth up and smite thee."
—old aviation proverb

CHAPTER 8

Shutting Down

*"...it will help all pilots to remember that profes-
sionalism is the key to overcoming all obstacles to
safety. We begin with the proper attitude, sustain
efforts toward excellence throughout each flight,
and strive for the highest possible human performance at all times. That is not only the key to achieve-
ment in aviation, it is the mark of a professional in
every field."*

—Donald D. Engen, former Navy pilot, former FAA Administrator
and current Director of the Smithsonian Institute's
National Air and Space Museum

At this point, you have completed a flight around the traffic pattern
and landed safely, but do not relax yet. There is an old saying that you
do not stop flying until the plane is tied down. This quip originated in the
days when everyone flew tail dragger aircraft. I have found that it is wise
to heed this advice regardless of what aircraft you fly, even to this day.

Also, do not be too quick about cleaning up the airplane from the
landing configuration before you clear the runway. Wait to clear the
flight environment, back across the hold-short line. When you are on the
runway centerline and decelerating to turn off the runway, look for the
yellow taxiway stripe to lead you clear of the runway, and then follow it
until you are across the hold-short line. Only then should you stop to
clean up the aircraft from its landing configuration. Waiting until this
point prevents you from being distracted. Distraction might cause you to

hit a taxiway light or accidentally move the landing gear knob when you intended to raise the flaps. The procedures and habits you develop now in a fixed gear airplane will carry through to upgrade aircraft later.

Once clear of the runway, announce on the radio that you have cleared it so that the runway is free for following traffic. Place your hand on the flap switch or lever and say out loud, "Flaps identified, flaps up." Then raise the flaps. Take off the carburetor heat as soon as possible. Remember that this air is non-filtered and, if left on, harmful dust and debris can enter the internal workings of the engine through the carburetor. Turn off the landing light. It is not needed now (unless, of course, it is night) and turn off the transponder. On the Tomahawk, you can turn the electric fuel pump off as well. The pilot and passengers should keep their seatbelts fastened until the airplane comes to a complete stop at the tie-down location.

The same rules used for taxiing out for takeoff apply now for taxiing back to the ramp. Once back over your tie-down spot, you can employ a simple memory aid to help with the proper sequence for shutting down. I call it the 5 M's: Music, Mixture, Mags, Master, Maps.

- ✈ The **Music** consists of turning off all the radios, lights, and electrical switches except the master.
- ✈ The **Mixture** should be closed completely to starve the engine of fuel.
- ✈ The **Magnetos** should be turned off only after the propeller has stopped, thus burning all fuel.
- ✈ The **Master** switch now can be turned off to disconnect the battery circuit.
- ✈ The **Maps** are the papers that you must gather to take with you, such as Hobbs reading, sectionals and the like.

This memory aid has never failed me, even while flying light twin engine aircraft. Finally, you must be certain to physically secure the airplane. This involves control locks, chocks, tie-downs, and everything else contained on your checklist. One thing not found on any checklist I have seen is the practice of closing all open, outside air vents. This helps prevent insects from entering and visiting you during your next flight. If you must leave air vents open, install screen material over or in them to keep multi-legged visitors from the cockpit.

Frequently I have been surprised at the number of pilots who do not know how to tie a good aviation knot. At one time or another, most pilots have been forced to try to untie someone else's particularly tough knot. If the knot were like some I have encountered, those pilots were most likely tempted to simply cut it with a pocket knife. There is one particular knot that I use which is simple, strong and does not leave any slack in the rope. Most important, it is easy for the next person to untie (see Figure 8-1).

To tie this knot, first push the end of the rope through the tie-down loop attached to the airplane. Then pull the rope tight with one hand. Continue to hold the rope tight with that hand, and use your free hand to loop the loose end of the rope around the tight part of the rope as you would to make a half hitch knot. Pull this second hand down to pull the

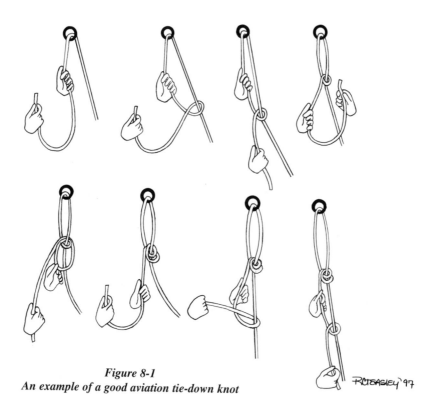

Figure 8-1
An example of a good aviation tie-down knot

half hitch tight. Only now should you let go with the first hand. The first hand can now replace the second hand with no loss of rope taughtness. With the second hand, take the free end of the rope again and loop it back around and through where the first loop did. While continuing to hold the rope tight with the first hand, here comes the secret: instead of making two half hitch loops, position the second loop now at the back (or bottom) side of the first loop and pull down. The knot should bind on itself and will hold taughtness in the rope as you let go with both hands. All that is left is to add another single half hitch safety loop below the binding knot and you have a strong, user friendly tie-down.

> *"Science, freedom, beauty, adventure—*
> *aviation offers it all."*
>
> —Charles Lindbergh

CHAPTER 9

Imminent Situations

"Always leave yourself a way out."
—Chuck Yeager, former Test Pilot

The dictionary describes the word *imminent* as "ready to take place; hanging threateningly over one's head; likely to occur at any moment." Imminent situations are always a possibility with flying machines. Anything may occur at any moment, and we, as pilots, should be ready for anything. We never should take for granted any takeoff, enroute navigation, or landing. In fact, once you learn the basics of flight, only practice and further instruction will allow you to improve your skills. But good initial training will realistically simulate the stress and evaluate your ability to deal with the very real possibility of in-flight problems.

Coping with stress gets easier with practice. If you try to balance cups and saucers, one stacked on top of the other, you may be able to handle three or four. If the fifth and sixth cup and saucer are added too soon, you may drop the stack. With practice to build familiarity with the extra balance and added stress, however, it could be possible to work up to nine and ten cups and saucers on the stack. A certain amount of stress is good for a pilot. It keeps one sharp and helps prevent complacency. Too much stress too soon, however, can be a barricade to good decision making. The goal of a safe pilot is to continue improving his or her skills, knowledge, mental attitude, and decision-making abilities to be better prepared for the unknown and unforeseen stressful scenarios of flight.

Two events that I will discuss briefly are the aborted takeoff (also known as a rejected takeoff) and the go-around (also known as a rejected landing). While the POH describes procedures for handling these two situations and more, you will never have time to retrieve the POH and read it when a problem arises. The best thing to do is read in advance the "Emergency Procedures" section of the POH and commit all the immediate action procedures to memory. After immediate steps are taken, you usually can refer to the checklist to ensure that all necessary tasks have been completed. The three basic rules applying to all airplanes in all "emergencies" are:

1. Maintain control of the aircraft—you must remain a pilot at all times, otherwise you are simply a passenger.

2. Diagnose the situation and take proper action(s)—you must stay calm, handle sudden stress, and think.

3. Land as soon as necessary—you must decide whether to land at an immediate location or continue to an airport.

There are four basic scenarios during which you should abort a takeoff. One is during the takeoff run when you are not yet airborne. Another is immediately after leaving the ground while there is still a safety margin to land straight ahead back on the runway. Still another, and probably the worst scenario, is when you are too far into the takeoff to land back on the runway (straight ahead), yet you are at too low an altitude to turn and land on the runway in the opposite direction from which you just departed (see Figure 9-1). The final scenario assumes that you have gained sufficient altitude to make a 180° (then plus some) turn and land back on the runway in the opposite direction from which you just departed. All of these scenarios represent unexpected events and require discipline, knowledge, and immediate action.

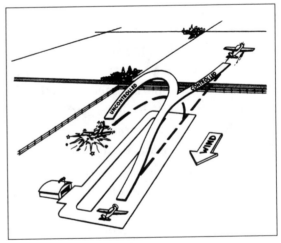

Figure 9-1
When aborting a takeoff, sometimes the best option is to land straight ahead

The time to begin your emergency training and thought process is before takeoff. Remind yourself that these situations can happen to you. Avoid denial at all costs and have a back-up plan ready. Never assume that a takeoff is guaranteed. Be ready to abort the flight at the first moment of trouble. By itself, this mental attitude will quicken your reaction time and could be the difference between good success and marginal success. The FAA states that, on average, there are 4 to 5 seconds between problem recognition and pilot action. This may not sound like significant reaction time, but you must remember that even at 59 KIAS, you use up runway at about 100 feet per second.

Know your airplane, its procedures, performance, and systems backward and forward. As much as I would like to, there is no way that I can describe every possible scenario here. There are too many variables among aircraft, runways, density altitude, winds, and other factors. Each time a problem occurs, it is unique to that pilot at that time. He or she alone must decide the correct course of action based on his or her knowledge of the aircraft and the conditions at hand.

Build in your own safety margins. You should calculate the takeoff run and the landing distance required for the existing conditions using your POH. When you add these two together, you have a good indication as to how much runway length may be needed for a successful abort if you are at rotation speed. To be safer, take that number and add at least an additional 20%. This information now becomes a foundation upon which to make further decisions. By all means, use all the runway for takeoff instead of cutting off the corner after crossing the hold-short line. This is cheap insurance once again, because any runway left behind does you absolutely no good. In addition, be mentally ready to abort a planned flight during the engine run-up phase. It is better for you to taxi back to the ramp at this time rather than have someone else truck you back later.

Sometimes you, as the pilot, may make a decision to abort the takeoff because something does not feel, sound, or smell right. At other times, the airplane's engine may make the decision for you. Either way, the procedures are basically the same. Once you have made the decision to abort, do not second guess yourself or deny what is happening to you.

Discipline must kick in, and you must act quickly with good decisions and instant actions. Some possible aborted takeoff scenarios and recommended actions are as follows:

Example 1: Aborted Takeoff—Not Airborne

In this situation, you should:

✈ Close the throttle immediately, completely, and without hesitation. Every instructor will tell you that this is the first thing to do. There are no options.

✈ Retract flaps that have been added for the takeoff. This action will reduce some lift and allow more braking force.

✈ Ease and hold the control yoke full aft. This procedure will keep the nose wheel light and prevent wheelbarrowing, while increasing the down load on the brakes to make them more effective.

✈ Apply full brakes, but not to the point of locking the brakes and skidding the tires. This type of skidding is known as reverted rubber hydroplaning and will not result in effective braking.

✈ Remember that some of the procedures above can be performed concurrently.

Example 2: Aborted Takeoff — Airborne — Runway Remaining

In this situation, you should:

✈ Close the throttle immediately and lower the nose.

✈ Maintain airspeed above stall speed and land straight ahead. Do not retract flaps while in flight.

✈ Once on the ground, retract flaps, ease and hold the control yoke full aft, and apply full braking action as necessary.

Example 3: Aborted Takeoff — Airborne — No Runway Remaining

In this situation, you should:

✈ Close the throttle immediately unless there is sufficient power remaining to limp back to the runway. This is a critical decision time. (If the throttle is closed, attain best glide speed and maintain aircraft control.)

✈ Pick your spot to land quickly. Most instructors agree that straight ahead is preferable if there are no obstructions. Trying to make a 180° turn is not wise if sufficient altitude has not been gained. The altitude you have gained will determine how much time you have and how far you can glide.

✈ Ensure that all fuel and electrical switches are off before touchdown.

✈ Add flaps as necessary and increase to full flaps at some point before touchdown. Crash as slowly as possible.

Example 4: Aborted Takeoff — 180 Degree (Plus) Return to Runway

The first thing that you must realize about the "impossible turn" is that it really requires more than 180° of turn because the turn itself displaces you a certain distance from the runway centerline. The extra maneuvering required to make it back to the centerline may actually add up to 360°. You can increase your odds, however, by making the right decisions for the conditions at hand. For simplicity, I will refer to it as the 180° turn.

In case of an aborted takeoff where a 180° return to the runway is chosen, the first thing to remember is there is a very good reason why this maneuver is sometimes called the impossible turn. Indeed, there are too many accident reports, many of these fatal, that describe a pilot attempting this turn after his or her engine failed on takeoff. The pilot entered a stall, or, more disastrous, a spin at low altitude and never recovered. But neither is landing straight ahead or off to the side always the best decision. Altitude is the single factor in this equation that determines success or failure. If sufficient altitude has been gained, a 180° turn may be the best decision.

The best way to gain altitude for time after takeoff is to attain Vy as soon as possible and hold it. In order to know how much altitude is enough to successfully accomplish a 180° turn, perform a simple test with your airplane. The test is: climb to a safe altitude and hold it, retard the throttle and hold that altitude until best glide speed is attained. Then make a 360° turn at best glide and upon roll out, wings level, back to the original heading, note the altitude lost. This will be a rough indication of the altitude necessary to execute a 180° turn back to the runway for a landing.

The decision to turn right or left should not be predicated on how comfortable you are when turning one way or the other. Rather, if there is a crosswind, turning into the wind can decrease your lateral displacement from the threshold. Also, the angle of bank will affect the result. Too shallow a bank will result in too slow a turn rate, thereby actually increasing the altitude lost in the turn as well as the lateral displacement. Obviously you should not increase the bank excessively because the stall speed and load factor increase with the increase in bank. A compromise is the best decision, and that usually is about 40° to 45° of bank at best glide speed, directly into the face of any crosswind.

The last thing to remember about the "impossible turn" is that you will be landing with a tailwind component. You must adjust your approach accordingly and perhaps add flaps sooner than you would under normal conditions. Do not discount the possibility that you can land on a crossing runway, parallel runway, or even a taxiway. If another runway is your best option, a radio call should be made as soon as possible to warn other aircraft of your situation and intent. There are no guarantees with the 180° "impossible turn." If you are not comfortable with this maneuver or are in doubt of successfully completing it, do not attempt it. But it can be done with the right preparation, decision making, and skill level. The decision is yours. No matter what your course of action, the most important factor is to have an option plan before you attempt a takeoff.

Go-Around (Rejected Landing)

Just as you should never expect a takeoff to be assured, never expect any landing to be assured either. Occasionally, it may be wise to abandon a landing approach. This decision is based on safety considerations and the mental comfort level of the pilot. If a landing does not feel right, it probably is not right. Situations that might warrant a go-around decision include, but are not limited to, extremely low base-to-final turns, overshooting the runway centerline, appearance of unexpected hazards on the runway, wake turbulence from preceding aircraft, overtaking another aircraft on final approach, too slow an airspeed while already positioned lower than normal on an approach, drifting sideways while landing in a crosswind, and a bounced landing.

It is not disgraceful or embarrassing to make a go-around. It is a wise decision considering the consequences that might result should a needed go-around not be executed. The decision to make a go-around can be made anywhere in the landing process, although the most critical time to make the decision is while very close to the ground. Overall, the best time to reject a landing is the first time the thought pops into your head. Hesitation can create problems. The earlier a situation is recognized and a go-around is initiated, the safer the procedure will be. It is imperative that a good knowledge of go-around procedures for your aircraft be committed to memory and implemented as necessary.

The first action, once you have made the decision to go-around, is to add full available power immediately. But never jam the throttle forward aggressively. Smooth and steady application of power is all that is necessary. In some cases, jamming the throttle forward too fast is detrimental. Remember in the earlier chapter on engine starting when we talked about accelerator pumps on carburetors? Now is the time when an accelerator pump is your best friend. For the C-152, which has an accelerator pump, the smooth but brisk application of power is usually more forgiving because the accelerator pump will give the engine the fuel needed to mix with the sudden rush of air that the throttle also provides. In the Tomahawk, however, where there is usually no accelerator pump on the carburetor, the sudden and jamming application of power can cause the engine to sputter—even quit—because the sudden air input from the throttle exceeds the proper amount of fuel/air mixture that the engine can handle. The key is to react immediately, but with a smooth and steady motion.

When full power is applied, right rudder usually will be needed to counteract torque, spiraling slipstream, and other factors. The next action depends on whether full flaps have been applied. If they have been applied fully prior to the go-around decision, retract one notch of flaps for the second step. Full flaps add tremendous drag and may prevent the airplane from climbing. Once flaps have been reduced by one notch to less than full, attain the climb speed recommended for your airplane to suit the existing conditions. For example, if you have initiated the go-around close to the runway and the runway has trees at the far end, then the best angle of climb (Vx) might be appropriate for a short time to clear the trees. Otherwise, Vy is usually the best speed for a go-around. Under no circumstances should you attain a speed less than Vx, nor should you retract all the flaps at once.

The go-around is a maximum performance maneuver that is executed in sequential steps to assure success. Upon attaining full power and after the first notch of flaps is retracted, glance at the VSI for a climbing trend. When an initial climb is seen, all obstacles are cleared, and after Vy is attained, retract a second notch of flaps (usually it is not wise to retract a second notch of flaps at Vx). Once again, glance at the VSI to ensure a continuing climb, then retract another notch of flaps. The progression continues until all the flaps are fully retracted and a safe airspeed and altitude are attained. From this point, the climb-out becomes relatively normal.

During landings, be cautious about applying too much nose-up trim during the final approach. This leads to a sharp nose-up movement when power is applied for the go-around. The result would most likely be that the aircraft reaction will distract the pilot from implementing proper procedures as he or she struggles to readjust trim. The ever present prospect of a go-around is why I do not recommend adjusting the trim to full nose-up, and I do advocate waiting until the field is assured before adding full flaps. Should a go-around be required sooner in an approach, the pilot workload is reduced by not having to deal with a high drag, lurching airplane situation.

A pilot learning to fly usually is conditioned to respond to the command "Go around" when issued by the instructor. With that command, the student learns to react appropriately and correctly with the procedures for a go-around. This conditioned response works well for the student, the instructor, and for air traffic control, because they recognize this as a command/response. But what about the real world, the world in which a non-aviator is in the cockpit with a pilot? Passengers are not trained to elicit the proper command of "Go around!" As an instructor, I try to simulate the real world as much as is safe and feasible.

Imagine a pilot landing with his non-pilot mother sitting beside him in the right seat when, on short final, a dog saunters onto the runway. The dog is blissfully unaware of imminent danger and does not hurry his pace. Perhaps it finally senses the danger and freezes on the runway. Meanwhile, the pilot is carrying out his duties, checking the cockpit to ensure that all is prepared for the landing, chatting with his mom, and feeling excited but a little nervous about her flying with him. In fact, he is so careful inside the cockpit that he never sees the dog. Suddenly, his mom shouts, "There's a dog on the runway!" He looks up, sees the dog, hesitates in deciding what to do, and is unable to prevent the plane from striking the dog.

The problem here is multi-faceted, but let us concentrate on the aspect of reaction time. Because the pilot did not hear a standard or familiar command, it took him longer to assess the situation before reacting. This delay led to an untimely and messy end for the dog, not to mention likely damage to the airplane and mental trauma for the passenger. On occasion, my pro-active instructional style for this potential real life situation has been to give the student a non-standard exclamation on the short final approach such as "There's a deer on the runway!" and then see how he or she handles the decision-making process. I get to see how, and how quickly, the student reacts; and the student gets to practice handling the stresses of "normal" flying as well.

Interestingly, two imminent situations occurred, cool heads prevailed, and finally, a go-around decision was made, even on the very first manned flight in recorded history. It was made in a non-captive, lighter-than-air balloon. The balloon was made by two brothers, Joseph and Etienne Montgolfier, and was fueled by a wood fire on a grate built beneath the balloon's throat. The fire produced the heated air needed for lighter-than-air lift. The takeoff took place just outside of Paris. The date was November 21, 1783. The pilot was Jean-Françios Pilâtre de Rozier, a young physician and scientist, and with him was his friend, François Laurent, the Marquis d'Arlandes, an infantry major. They flew for over twenty minutes, covered more than five miles, and reached an estimated altitude of 500 feet above the roof tops of Paris. Later, the Marquis wrote a letter to one of his friends describing the incident:

> "I heard a new noise in the machine, which I thought came from the breaking of a cord. I looked in and saw that the southern part was full of round holes, several of them large. I said: 'We must get down.'
>
> "'Why?' (asked Pilâtre). 'Look!' said I.
>
> "At the same time I took my sponge and easily extinguished the fire which was enlarging such of the holes as I could reach, but . . . I repeated to my companion: 'We must descend.' He looked around him and said, 'We are over Paris.'
>
> "Having looked to the safety of the cords, I said: 'We can cross Paris.'
>
> "We were now coming near the roofs. We raised the fire and rose again with great ease."

Although they landed safely, you can see that even on this very first free flight in the history of man, an emergency situation was averted (with a wet sponge), a decision was made, power was added, and a go-around was exercised as an alternative to crashing into the roof tops of Paris. Even today, we as pilots can do no less. The number one rule is to stay calm and fly the aircraft.

*"We're all accident prone. Flying does present hazards.
If your emergency training is up to date,
you can survive an emergency."*

—Tony LeVier, former Test Pilot

CHAPTER 10

Instructor's Thoughts

"The trick, Fletcher, is that we are trying to overcome our limitations in order, patiently. We don't tackle flying through rock until a little later in the program."

—Jonathan Livingston Seagull, Flight Instructor,
in *Jonathan Livingston Seagull: A Story* ©1970 Richard Bach

Touch-and-Go Landings

I have very strong feelings regarding the practice of touch-and-go landings. I feel that they are accidents waiting to happen. Too many times have I heard the story about the pilot who retracted his or her landing gear on roll-out, thinking that he or she was retracting the flaps, or that a pilot lost directional control because of attention distraction during a crosswind landing. Regardless of what certificate or rating you have, the roll-out is no place to be distracted from the absolute necessity of controlling the airplane.

There is good reason that it is a practice of many corporate flight departments, as well as FAR Part 135 single pilot charter operations, not to touch anything on the instrument panel until the aircraft is clear of the runway. It is interesting to note that while this rule is not written in the FARs specifically, it is written into the operations manuals that the pilots have to read and sign. Note also that this practice is common for single pilot operations where the pilot is very busy with complex equipment.

Two-pilot operations may or may not mandate this policy due to the shared workload.

The practice of touch-and-go landings usually is incorporated into the student pilot's regimen of training. I believe that this is the place where the practice first should be discouraged. After all, are not all instructors trying to instill good operating practices and procedures into their students? The practice of hurrying to re-configure the airplane for an immediate rolling takeoff is never a good one, especially for student pilots. Students are extremely impressionable, and they are likely to believe that any practice an instructor shows them is appropriate.

It is a much better teaching practice to exit the runway, use the checklist to re-configure the airplane, and then taxi on the taxiway back to the runway. At the least, if you are on a long runway and there is no traffic behind you, come to a complete stop and then re-configure. This is called a stop-and-go. But even this practice is less desirable because now you are taking off with useless runway behind you. Stop yourself and think about your next move and the possible problems which may arise.

In the end, however, it is the instructor's sole discretion whether or not to teach touch-and-go landings. It also is the instructor's discretion to decide whether or not a student is overloaded and needs a full stop. Surely the practice of full stop, taxi back landings will add some additional cost to the student's private pilot certificate course. It also will put a little more time in the student's and instructor's logbooks. While I do not advocate the use of full stop, taxi back landings as a means for an instructor to build time towards a flight career, I do endorse that this knowledge of options be imparted to the student within his or her flight training regimen and that good communication remain paramount in the student/instructor relationship.

A similar issue to consider is the short field landing. It is a well known fact that raising flaps during maximum performance short field landings will increase the download on the main wheels, thereby increasing braking effectiveness. My experience shows that this action may decrease the landing roll somewhat, but it will not make a significant difference overall. In my opinion, the large gain in employing good landing procedures outweighs the minimal gain in decreased landing distance, especially in a trainer. In any case, consult your POH for the proper technique for your aircraft.

If you absolutely must retract the flaps during a landing roll, one technique a PIC may consider is having another person in the right seat with the assigned task of operating the flap control at a specified or commanded time. This is like having another crewmember and working with cockpit resource management. Just be sure that any person assigned to a particular task understands exactly what is expected of him or her. Otherwise, you may end up paying for expensive prop, engine and/or airframe repair.

For the student pilot on a checkride, however, the best procedure I have found is having the student properly exit the runway as described before. Then the student should stop the aircraft and, in addition to his or her other duties, identify the flap control visually. Verification of the flap control by touch (but without moving the control lever or switch) is next, followed by a declaration that "flaps are identified." Only after the flap control is verified visually, and by touch, should it be moved to retract the flaps. This recommendation echoes one of my opening statements that a pilot's practices and procedures can be traced back to initial flight instruction. Instructors must be vigilant about where students place their hands. Visual verification of a control's movement is often paramount in a complex cockpit where levers, knobs, and switches abound.

Remember that there is a difference between technique and procedure. While techniques and procedures will vary among instructors, good procedures must take precedence over technique. Procedures may be regulatory, but techniques are not. Good, fundamental procedures learned from a conscientious primary instructor will stay with a pilot as he or she fine-tunes technique with subsequent instructors. Many times, good, solid, steady procedures and judgment will impress a pilot examiner more than the actual speed or technique with which the student can perform the necessary tasks. Always refer to your "After Landing" checklist, however, for the specific sequence to reconfigure your aircraft.

The Pilot Proficiency Award Program

After a pilot certificate is obtained, regular proficiency training is essential to the safety of all pilots. The Pilot Proficiency Award Program, sometimes called the "Wings" program, is designed to provide pilots with the opportunity to establish and participate in a personal recurrent training program. Completion of a phase of the program also counts as a biennial flight review (BFR) as stated in FAR 61.56. In most cases, though, participation in a phase of the program will require going beyond the minimum requirements for a BFR. However, this is a program that acknowledges pilots who are safety conscious enough to go beyond the minimum requirements for currency. Sometimes maligned and generally underused, this program remains one of the best programs ever instituted by the FAA for general aviation (GA).

There are 20 phases to the program and each pilot completing the requirements of the first 10 phases will receive a "Wings" pin and a certificate. A pilot can receive only one "Wings" pin per year. For phases 11 through 20, the participating pilot will receive a certificate only. All pilots holding a recreational or higher pilot certificate and a current medical certificate, when required, may participate. There is a flight regimen for airplanes, seaplanes, rotorcraft, gliders, lighter-than-air craft, and even ultralights. Successful completion of a mountain flying course will also serve to satisfy the training requirements. In addition to the flight requirements necessary to complete a particular phase of "Wings," you must attend an FAA sponsored or FAA sanctioned safety seminar. Advisory Circular 61-91 explains the Pilot Proficiency Award Program in greater detail.

I once overheard a pilot stating to another pilot in reference to the FAA "Wings" program, "What would I want those cheap plastic wings for?" Well, for one thing, they are not cheap plastic wings. They are a set of metal, tie-tack sized, wing-shaped pins that look distinctive and sharp on a tie or the lapel of a sport coat. Most pilots feel great pride in wearing them because they know that the winged pins cannot be bought or obtained without completing a phase of the award program. They are a badge of honor that must be earned. The certificate is suitable for framing, as well.

There are several other aspects of the program that are less known. Some aviation insurance companies offer a discount on premiums to clients who participate in the "Wings" program. They know that any pilot who makes the effort to go beyond the minimum recurrent flight training and attends a safety seminar is dedicated to safety and less likely to have an accident. Ask your insurance company whether it offers this incentive.

Another little known fact about the Pilot Proficiency Award Program is that a newly minted private pilot can immediately apply for phase 1 of the program if he or she has attended an FAA safety seminar and his or her instructor has signed off the requirements for phase 1. Check with your CFI, the appointed Aviation Safety Counselor in your area, or the Safety Program Manager at your nearest Flight Standards District Office. Get your "Wings," and wear them proudly.

Other reading

There are a multitude of aviation organizations and quality publications available to pilots for periodic reading. There is, however, no possible way that you can read all of them. Only you can decide which are best for you to join and/or subscribe to. Because it is imperative that you be aware of current changes, I suggest that you choose at least one since the aviation field is so dynamic. Furthermore, it helps to read good proficiency articles as often as possible. What many pilots do not know, however, is that there are a couple of good newsletters with no-cost subscriptions.

One of these free publications is NASA's "CALLBACK" newsletter. An offshoot of the Aviation Safety Reporting System (ASRS), this publication reprints edited versions, with comments, of the best and most relevant of the submitted NASA safety reports. The information contained in this monthly newsletter is current, pertinent, derived from real life situations, and there is no advertising. Write to them to request a NASA safety report form and ask to be added to the "CALLBACK" newsletter mailing list. The address is:

Aviation Safety Reporting System
Ames Research Center
P.O. Box 189
Moffett Field, CA 94035-0189
Phone: (415) 969-3969

You also can reach them on the Internet at http://olias.arc.nasa.gov/asrs.

Another free newsletter comes from the United States Aircraft Insurance Group (USAIG). This bi-monthly newsletter usually reprints one or two good articles on aviation topics from other magazines. They probably prefer to have aircraft owners or potential aircraft owners (and customers) as their mailing base, but the information is excellent and there is only one advertiser. Write, call, or fax to:

USAIG
One Seaport Plaza
199 Water Street
New York, NY 10038
Phone: (212) 952-0100
Fax: (212) 809-7861

Another free resource is the US Government Printing Office's (GPO) Advisory Circular Checklist. Actually a catalog, this publication contains a listing of all GPO publications available and forms for ordering. Some of the publications require payment. These are appropriately noted with their associated fee, but many publications are free. There is even a mailing list, and you may request that your name be added to it to receive subsequent free advisory circulars. You will have to decide for yourself which of these publications to request when you receive your copy of the Advisory Circular Checklist. You can copy the enclosed order form (see Figure 10-1), and mail it or fax it.

On the Internet, the FAA's main server is available at

http://www.faa.gov. It hosts information from various FAA organizations. Advisory circulars also can be ordered on the web at http://www.faa.gov/abc/ac-chklst/actoc/htm. or you can try the GPO Internet site for ordering documents at http://www.access.gpo.gov/.

"The only man who never makes a mistake is the man who never does anything."

—Theodore Roosevelt, 26th U.S. President, who was in office during the Wright brothers' success on Thursday, December 17, 1903

CUT ALONG THIS LINE. COMPLETE ORDER AND MAIL TO SVC-121.23 LANDOVER, MD 20785

ORDER BLANK [Free Publications] **DATE____/____/____**

For Faster Service Use A Self-Addressed Mailing Label.

Mail To: U.S. Department of Transportation
Subsequent Distribution Office, SVC-121.23
Ardmore East Business Center
3341 Q 75th Ave.
Landover, MD 20785

Help Line: 301-322-4961 FAX REQUEST TO 301-386-5394 DOT Warehouse

NUMBER	TITLE	QUANTITY

SVC-121.23

Request Filled By: ______________________ **Date: ___/___/___**

1. Out of Stock [reorder in ____ days]**

2. Being revised

3. Cancelled, no replacement

4. Cancelled by ______________ [enclosed]

5. Other: ______________________

****IF YOU DO NOT RECEIVE DESIRED PUBLICATION(S) AFTER YOUR SECOND REQUEST, PLEASE CALL FAA'S TOLL-FREE CONSUMER HOTLINE: 1-800 FAA-SURE.**

TO COMPLETE ORDER **Enter Name and Address** **DO NOT DETACH**

NAME ______________________

STREET ADDRESS ______________________

CITY **STATE** **ZIP CODE** ______________________

Figure 10-1

CHAPTER 11

Closing Thoughts

"... And of the living ... none, not one
Who truly loves the sky
Would trade a hundred earth-bound hours
For one that he could fly. ..."

—Gill Robb Wilson, World War I pilot,
from his poem, *First Things First*

Some students call me a tough instructor. Some might use much stronger adjectives. This banter has never bothered me. I consider it sort of a compliment, because I consider myself an instructor who cares. I care enough to want to do it right. A smart student does not want an easy instructor who lets him or her slide by, barely meeting minimum requirements. Good training is sometimes tough training. Remember that no matter what surprises an instructor throws at you, your flying machine and Mother Nature can always throw something tougher and meaner. Good, thorough flight instruction better prepares you for the stresses of unusual and unexpected situations and moments.

Every pilot should fly every aircraft with a little bit of fear. Take nothing for granted. The student never should take good flight instruction for granted, and the instructor never should take even the exceptional student for granted. And no one should ever take an aircraft for granted. It is only a machine and machines break. The only brain the aircraft has is the pilot's. Even with all the instruments on the panel, the most important instrument in a flying machine is between the pilot's ears. The pilot must be ready at any moment to make a decision and take action. Saying "No" to a flight or turning around and going back while en route sometimes is the best action. The PIC is the ultimate and final decision-maker for any flight.

It is known that a person can learn to manage a flying machine. People learn to fly every day. If this person is you, then you know or will come to know that unspoken feeling of uniting with an aircraft such that you and the aircraft are one. You will make the transition from awkwardly getting into an intimidating machine to actually strapping it around you. The wings no longer will be the airplane's wings; they will be an extension of your body. The landing gear legs will not be the plane's; they will be your legs. It is indeed a special feeling to know that you can leave Earth's grip and return to it under complete control. To rise above all other earth-bound creatures, travel long distances, and then to return safely can be an experience unmatched by anything on this planet.

There may be risk associated with flight, but it has been said that the most dangerous part of flying involves the drive to the airport. Indeed, statistics indicate that the likelihood of personal injury in a car crash is greater than the probability of injury in a plane crash. We must accept risk in almost every kind of travel. Certainly, anytime you put yourself in a machine that moves across the land, on the water, or through the air, there is risk. If you are the one in control, however, your risk can be minimized through training, learning, and practice.

To commit to learn aviation is a commitment to a lifetime of learning. No two days are alike, no two aircraft are alike, indeed no two landings in the same aircraft on the same day are alike. Every time you rise into the air, you commit to learn and ultimately draw from your pool of learning experiences to get you and your passengers back on the ground safely.

Never dismiss the fact that you can learn from others. In fact, you must learn from others. From the time of the Wright brothers, there has been an exponential sharing of aviation knowledge. We are the product of that knowledge. We were not born with the genetics to fly like birds. Every one of us must learn from the beginning, from nothing, from others. Even the Wright brothers, who had no actual flight instructor available at the beginning of this century, had each other and a few Europeans from whom to extract knowledge and counsel.

In a letter to Octave Chanute dated 16 November 1900, Wilbur

Wright himself stated it this way:

"Now, there are two ways of learning how to ride a fractious horse: one is to get on him and learn by actual practice how each motion and trick may be best met; the other is to sit on a fence and watch the beast a while, and then retire to the house and at leisure figure out the best way of overcoming his jumps and kicks. The latter system is the safest; but the former, on the whole, turns out the larger proportion of good riders. It is very much the same in learning to ride a flying machine; if you are looking for perfect safety, you will do well to sit on a fence and watch the birds; but if you really wish to learn, you must mount a machine and become acquainted with its tricks by actual trial."

Thank you, Wilbur, Orville, and everyone else who has learned before me and subsequently built their bridges, passing along their knowledge, skills, and experiences across the chasm, deep and wide. An airplane truly unites our greatest tool—our hand, with our greatest dream—to fly. Ours is the world of man (or woman) and flying machine. What a wonderful combination. What a great journey. Learn it, do it, and pass it on.

"One can never pay in gratitude;
one can only pay in kind
somewhere else in life."

—Anne Morrow Lindbergh

Some Aviation Quips

Any day spent above ground is a good day.

Flying is the second greatest thrill known to man. Landing is the first.

If you can't afford to fly it right, be sure you can afford to fly it wrong.

In flying, flexible is too rigid; you have to be fluid.

Never fly a plane that doesn't have the paint worn off the rudder pedals.

Remember, you fly with your head, not your hands.

If an engine quits, fly it to the ground, don't fall it to the ground.

During a forced landing, fly the plane as far as possible into the crash.

Never let an airplane take you someplace your brain didn't get to five minutes earlier.

The three most useless things in aviation are the air above you, runway behind you, and fuel left on the ground.

The two most dangerous words in flying: "Watch this!"

The three most dangerous words in aviation: "It's the gauge."

Remember that an FSS forecast report is a horoscope with numbers.

Keep looking around; there's always something you've missed.

The only two times you can have too much fuel is when you're over weight or on fire.

It's far better to be on the ground wishing to be up there than up there wishing to be on the ground.

The purpose of the propeller is to keep the pilot cool. If you think not, stop the propeller and watch him sweat.

If something can go wrong, it will; even if it can't, it will.

Flying is a great life for men who want to feel like boys, but not for those who are.

The nicest VFR weather can be just as dangerous as the worst IFR weather.

A check ride should be like a pair of cut-off jeans—short enough to be comfortable, long enough to cover everything.

Your new certificate is only a permit to learn.

No airplane is impressed by the ratings on your ticket.

The time spent flying is not deducted from one's life span.

If you don't privately think you're the best in the game, you are probably in the wrong game. If you don't think you can make a mistake, you are definitely in the wrong game.

There are old pilots and bold pilots, but not many old bold pilots.

If the pilot survives, you will never learn what really happened.

This would be nice work if we didn't have to go on all these trips.

My new-hire copilot was so nervous when I told him to file VFR, he misspelled it.

The command, "Takeoff power," is not the same as "Take off power!"

Flying is hours upon hours of sheer boredom punctuated by brief moments of stark terror.

The landing is your signature at the end of any flight.

There are two kinds of pilots: those who have and those who will.

Flying is the hardest thing to learn and the easiest to do.

Flying is the hardest thing to teach and the easiest to do.

Ignorance in flying is the ever present danger.

References

Doud, Margery and Parsley, Cleo M. (Collected and arranged by). Father: An Anthology of Verse. E.P. Dutton and Co., Inc., New York. 1931. p.86.

Flight Training Handbook. US Department of Transportation, Federal Aviation Administration, Flight Standards Service. Advisory Circular 61-21A. Revised 1980.

Gum, S. and Walters B. An Invitation to Fly: Basics for the Private Pilot. Wadsworth Publishing, Belmont, CA. 1982.

Jensen, Paul (editor). The Fireside Book of Flying Stories. Simon & Schuster, New York. 1951. pp. 20-21.

Petersen N. Mixing winter flying with oil, fuel and water. Sport Aviation; December 1986, pp. 56-60.

Pilot's Handbook of Aeronautical Knowledge. US Department of Transportation, Federal Aviation Administration, Flight Standards National Field Office. Advisory Circular 61- 23B. Revised 1980.

Pilot's Operating Handbook. Cessna Model 152 (1978). Cessna Aircraft Company, Wichita, KS.

Piper Tomahawk Information Manual. Piper Aircraft Department. Issued 1978.

Private Pilot Manual. Jeppeson Sanderson Inc. Englewood, Co. 1988.

Wilson, Gill Robb. Leaves from an Old Log. Publisher unknown. Publishing date unknown.

IMAGINATION OF A THINKING CHILD

A child watches the clouds go by and he(((dreams)))in the misty-realm where shapes-merge and twist into being the reality that a

skywatcher should see to be inspired by that imaginary world.

WILLIAM   P   HEITMAN

Additional copies of

Flying & Learning: Basics For Every Pilot

are available by mail order for $14.95 per copy,
plus $3.00 for shipping and handling.

North Carolina residents please add an additional $0.90 for sales tax.

Please make checks and money orders payable to:

Dreamflyer Publications

Inquiries regarding quantity discounts to flight schools,
high schools, colleges, and universities are welcome.
Please submit request for discount on school
or institution letterhead.

Send all correspondence and orders to:

Dreamflyer Publications
P.O. Box 11583
Durham, NC 27703
E-mail: Drmflyrpub@aol.com